MASONRY VENEER

MV

Third Edition

MASONRY
INSTITUTE OF
AMERICA
www.masonryinstitute.org
(800) 221-4000

ICC
INTERNATIONAL
CODE COUNCIL®
INTERNATIONAL
CODE COUNCIL
www.iccsafe.org
(888) 422-7233

MASONRY VENEER

Third Edition

John Chrysler, P.E.
Executive Director
Masonry Institute of America

Thomas Escobar, Assoc. AIA
Design Director
Masonry Institute of America

Based on the previous editions developed by:

James E. Amrhein, S.E.

Published by

MASONRY INSTITUTE OF AMERICA
(800) 221-4000
www.masonryinstitute.org

INTERNATIONAL CODE COUNCIL
500 New Jersey Avenue, NW, 6th Floor
Washington, DC 20001-2070
www.iccsafe.org
(888) 422-7233

Third Edition
First Printing, March, 2008

ISBN-10: 0-940116-50-2
ISBN-13: 978-0-940116-50-4
Library of Congress Cataloging-in-Publication Data
Printed in the United States of America

This book was prepared in keeping with current information and practice for the present state of the art of masonry design and construction.

The author, publisher and all organizations and individuals who have contributed to this book cannot assume or accept any responsibility or liability, including liability for negligence, for errors or oversights in this data and information and in the use of such information.

MIA 505-08 03-08 2M

TABLE OF CONTENTS

PREFACE---vii

SECTION 1 INTRODUCTION---1

1.1 History of Masonry Veneer---1

SECTION 2 MATERIALS---3

2.1 General---3
2.2 Masonry Units---3
2.2.1 Clay Masonry Units---3
2.2.1.1 Grade of Brick---4
2.2.1.2 Visual Inspection---5
2.2.2 Concrete Masonry Units---6
2.2.2.1 Compression Strength---6
2.2.3 Stone---7
2.2.3.1 ASTM Stone Specifications---8
2.2.3.2 Classification of Stone---8
2.2.3.2.1 Geological Classification---8
2.2.3.2.2 Texture of Quarried Stone---9
2.2.3.2.3 Classification of Fabricated Stone for Soundness---10
2.2.4 Terra Cotta---11
2.2.4.1 Characteristics---11
2.3 Mortar---12
2.3.1 Proportion Specification---13
2.3.2 Mortar Materials---13
2.3.2.1 Cements---13
2.3.2.1.1 Portland Cement---13
2.3.2.1.2 Masonry Cements---13
2.3.2.1.3 Mortar Cements---13
2.3.2.2 Hydrated Lime---14
2.3.2.3 Mortar Sand---14
2.3.2.4 Water---14
2.3.2.5 Admixtures---14
2.3.2.6 Color---15
2.3.3 Pre-Blended Mortar---15
2.4 Types of Mortar Joints---15
2.5 Structural Support Backup Materials---17
2.5.1 Structural Masonry Backup---17
2.5.2 Structural Concrete Backup---17
2.5.3 Wood Stud Backup---18
2.5.4 Steel Stud Backup---18
2.6 Shelf Angles---19
2.7 Flashing---20
2.8 Weep Holes---22
2.9 Connectors---24
2.9.1 Wall Ties---24
2.9.2 Anchors---25
2.10 Expansion Joints---26

SECTION 3 SYSTEM DETAIL REQUIREMENTS---27

3.1 General---27
3.2 Veneer Requirements---27
3.2.1 Definitions---28
3.2.2 Installation---28
3.3 Adhered Veneer---29
3.3.1 Thin Brick---30
3.3.2 Honeycomb Stone---31
3.3.3 Terra Cotta---31
3.3.4 Adhered Veneer Installation---32
3.4 Anchored Veneer---34
3.4.1 Stone Veneer---37
3.5 Terra Cotta---42
3.6 Block Veneer, Concrete Units---43

SECTION 4 TYPICAL DETAILS---45

4.1 General---45
4.2 Roof/Parapet Details---45
4.3 Shelf Angle/Flashing Details---48
4.4 Sill and Jamb Details---51
4.5 Floor Connection Details---54
4.6 Wall Base Details---55
4.7 Expansion Joint Details---59
4.8 System Configuration---61

SECTION 5 GUIDE SPECIFICATIONS---63

5.1 Guide Specification for Adhered Thin Clay Brick or Stone Slab Masonry Veneer---63
5.2 Guide Specification for Adhered Veneer Using Latex Portland Cement Mortar---69
5.3 Guide Specification for Anchored Brick Veneer---72
5.4 Guide Specification for Exterior Stone Cladding---77
5.5 Guide Specification for Ceramic Veneer---89
5.6 Guide Specification for Furnishing and Installing Ceramic Veneer---91
5.7 Guide Specification for Veneer Tile Installed on Exteriors in Portland Cement Mortar---94

SECTION 6 GLOSSARY OF TERMS---97

SECTION 7 REFERENCES---117

SECTION 8 INDEX---119

PREFACE

Designers have historically chosen masonry veneer to adorn churches, cathedrals and monuments because of its beauty, economy and durability.

There are numerous methods that have been developed to secure masonry veneer to a structural backup. Building codes have adopted some of the better and safer systems as they evolved and became standardized. Currently most building codes recognize two basic methods to install masonry veneer. The first of these methods, called adhered veneer, secures the masonry units to the structural back-up using a bonding material. The second method, anchored veneer, attaches the masonry to the structural back-up using mechanical fasteners called anchor ties or veneer ties.

These two systems are outlined in the building codes with prescriptive requirements that allows their use with little engineering design. However, the International Building Code also allows the use of alternative systems based on engineering principles such as reinforced veneer, masonry panels and stone panels on structural frame back-ups.

The success of any veneer system depends on the proper design, selection of durable materials and quality construction.

ACKNOWLEDGEMENTS

The authors would especially like to acknowledge the contributions of James Amrhein, S.E. and Phillip Samblanet, P.E. for their work in the previous edition.

The contributions of Luis Dominguez and Debby Chrysler, MIA staff, were invaluable in the development and production of this publication.

Finally we wish to thank the Board of Trustees of the Masonry Institute of America for their constant support: Ron Bennett, Chairman, Doug Williams, Frank Smith, Ken Tejeda, Bobby Williams, Jim Smith, Dana Kemp and Julie Salazar who have given their full cooperation to see that this publication has been successful and a benefit for the masonry industry.

THE MASONRY INSTITUTE OF AMERICA

The Masonry Institute of America, founded in 1957 under the name of Masonry Research, is a promotional and technical research organization established to improve and extend the use of masonry. The Masonry Institute of America is supported by the California mason contractors through labor management contracts between the unions and contractors.

The Masonry Institute of America is active in California promoting new ideas and masonry work, improving national and local building codes, conducting research projects, presenting design, construction and inspection seminars and publishing technical and non-technical papers, all for the purpose of improving the masonry industry.

The Masonry Institute of America does not engage in the practice of architectural or engineering design or construction nor does it sell masonry materials.

INTERNATIONAL CODE COUNCIL®

Since the early 1900's, the United States had been served by three sets of building codes developed by three separate model code groups: Building Officials and Code Administrators International, Inc. (BOCA), International Conference of Building Officials (ICBO), and Southern Building Code Congress International, Inc. (SBCCI). These codes were extremely effective and responsive to regional needs. But, in 1994, recognizing the urgent need for a single set of codes that would serve national needs, the three groups united to form the International Code Council® (ICC®) with the express purpose of creating and developing one master set of comprehensive, coordinated, design and construction codes.

Substantial advantages are inherent to this single set of codes. Code enforcement officials, architects, engineers, designers, and contractors throughout the United States can now work with a consistent set of requirements. States and localities that currently write their own codes or amend the early model codes may choose to adopt the International Codes without technical amendments, which encourages consistent code enforcement and higher quality construction. Enhanced membership services are an additional benefit. All issues and concerns of a regulatory nature now have a single forum for discussion, consideration, and resolution. Whether the concern is disaster mitigation, energy conservation, accessibility, innovative technology, or fire protection, the ICC offers a means of focusing national and international attention on these concerns.

The ICC makes available an impressive inventory of International Codes™, including:

- *International Building Code®*
- *International Energy Conservation Code®*
- *International Fire Code®*
- *International Fuel Gas Code®*
- *International Mechanical Code®*
- *International Plumbing Code®*
- *International Private Sewage Disposal Code®*

- *International Property Maintenance Code®*
- *International Residential Code® for One- and Two-Family Dwellings*
- *International Zoning Code®*
- *ICC Performance Code for Buildings and Facilities™*
- *International Existing Building Code™*
- *International Wildland-Urban Interface Code™*

These codes provide a comprehensive package for adoption and use in the 21st Century.

The ICC also offers unmatched technical, educational, and informational products and services in support of the International Codes, with more than 300 highly qualified staff members at 16 offices throughout the United States and Latin America. Products and services readily available to code users include:

- Code application assistance
- Education programs
- Certification programs
- Technical handbooks and workbooks
- Plan reviews
- Automated products
- Monthly magazines and newsletters
- Publication of proposed code changes
- Training and informational videos

MASONRY STANDARDS JOINT COMMITTEE

The Masonry Standards Joint Committee (MSJC) is an organization comprised of volunteers who through background, use, and education have established experience in the manufacturing of masonry units and materials and the design and construction of masonry structures.

Working under its three sponsoring organizations, The Masonry Society (TMS), the American Concrete Institute (ACI) and the American Society of Civil Engineers (ASCE) the Committee has been charged with developing and maintaining consensus standards suitable for adoption into model building codes. Since The Masonry Society has received ANSI accreditation, TMS has become the lead sponsor in the production of the MSJC.

In the pursuit of its goals, committee activities include:

1. Evaluate and ballot proposed changes to existing standards of the Committee.
2. Develop and ballot new standards for masonry.
3. Resolve negative votes from ballot items.
4. Identify areas of needed research.
5. Monitor international standards.

In this publication the term 'MSJC Code' refers to Building Code Requirements for Masonry Structures (ACI 530/ASCE 5/TMS 402) and the term 'MSJC Specification' refers to Specification for Masonry Structures (ACI 530.1/ASCE 6/TMS 602).

THE MASONRY SOCIETY

The Masonry Society (TMS) founded in 1977, is an international gathering of people interested in masonry. It is a professional, technical, and educational association dedicated to the advancement of knowledge of masonry. TMS members are design engineers, architects, builders, researchers, educators, building officials, material suppliers, manufacturers, and others who want to contribute to and benefit from the global pool of knowledge on masonry.

AMERICAN CONCRETE INSTITUTE

The American Concrete Institute (ACI) is a technical and educational society founded in 1904 with 30,000 members and 93 chapters in 30 countries.

As ACI moves into its second century of progress through knowledge, it has retained the same basic mission: develop, share, and disseminate the knowledge and information needed to utilized concrete to its fullest potential.

AMERICAN SOCIETY OF CIVIL ENGINEERS

The American Society of Civil Engineers (ASCE) was founded in 1852 and currently represents 125,000 members of the civil engineering profession worldwide. ASCE's vision is to position engineers as industry leaders building a better quality of life.

To provide essential value to members, their careers, partners and the public, ASCE develops leadership, advances technology, advocates lifelong learning, and promotes the profession.

1 INTRODUCTION

1.1 HISTORY OF MASONRY VENEER

2006 Random House Webster's Unabridged Dictionary, © 2006 provides eight definitions for the word veneer. Three of these definitions are especially appropriate to the discussion of masonry veneer:

"**3**. *Building Trades*. a facing of a certain material applied to a different one or to a type of construction not ordinarily associated with it, as a facing of brick applied to a frame house."

"**6**. to face or cover (an object) with a material that is more desirable as a surface material than the basic material of the object."

"**8**. to give a superficially valuable or pleasing appearance to."

Similarly, the 2006 International Building Code defines veneer in Section 1402.1 as follows:

"**VENEER** – A facing attached to a wall for the purpose of providing ornamentation, protection or insulation, but not counted as adding strength to the wall."

Masonry veneer is a system which may use clay brick, concrete masonry, stone or terra cotta to adorn, enclose and protect a building. The primary difference between masonry veneer and structural masonry is that veneer is designed to carry only its own vertical weight. The lateral loading, due to wind, seismic or other lateral loads, is carried by a system behind the veneer – a backup system. Lateral loads, such as wind and seismic loads, are transferred directly from the veneer into the structural backup. Structural masonry, by contrast, can be designed to carry both vertical and lateral (horizontal) loads.

Since veneer is not intended to resist lateral loads, the masonry veneer can generally be relatively thin compared to the structural wall.

Yet the veneer still provides both the beauty and durability of masonry. Additionally, the veneer also increases the fire resistance, noise resistance and water resistance of a non-veneered wall.

FIGURE 1.1 Egyptian Pyramid located in Giza constructed around 2500 B. C.

The use of veneer is by no means a new development. In fact, limestone veneer covered some of the pyramids in Egypt and pieces of it can still be seen at the top of the Great Pyramid, Cheops.

Additionally, architects have historically chosen masonry veneer to adorn churches, cathedrals and monuments because of its beauty, economy and durability.

There are numerous methods that have been developed to secure masonry veneer to a structural backup. Building codes have adopted some of the better and safer systems as they evolved and became standardized. Currently the 2006 IBC recognizes two basic methods to install masonry veneer. One method is called *adhered* veneer which secures the masonry units to the structural backup using a bonding material.

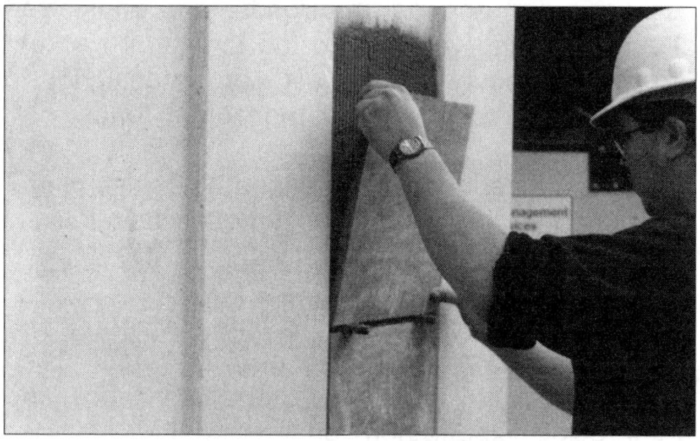

FIGURE 1.2 **Adhered veneer.**

The other method, *anchored* veneer, attaches the masonry to the structural backup using mechanical fasteners called anchor ties or veneer ties.

FIGURE 1.3 **Anchored veneer.**

These two systems are outlined in the building codes with prescriptive requirements that allow use with little engineering design. However, the code also allows the use of alternative systems based on engineering principles such as reinforced veneer, masonry panels and stone panels on a structural frame backup.

The success of any veneer system depends on the proper design, selection of durable materials and quality construction.

Details must be clear to ensure that a moisture-barrier veneer system will perform to prevent water intrusion into the building and a moisture drainage veneer system will effectively shed any anticipated water penetration.

Any anchored veneer system with an intentional airspace between the veneer facing and the structural backup is an example of a moisture drainage system. A quality veneer system will prevent the moisture from bridging the ties, thereby gravitating to the bottom of the system and evacuating through intentionally provided and functional weep holes.

In order to achieve a successful moisture drainage system, the components of flashing and weep holes must be present and care must be taken in the construction process in order to insure that these elements are functional. Moisture must have a clear passage to drain through the weep holes, which must be open for drainage of the moisture from the interior to the exterior of the veneer.

The corners, which are particularly susceptible to seismic damage, can perform very well if the designer understands wall movements relative to each other.

The anchored attachment system must have the flexibility of allowing vertical and horizontal movement relative to the backup system, and at the same time be capable of transferring the lateral wind and seismic loads to the backup system for ultimate ground distribution.

When properly designed and constructed, a masonry veneer system will survive natural elements, such as wind and earthquakes, with little maintenance required.

2 MATERIALS

2.1 GENERAL

Masonry veneer systems can be constructed from a variety of materials in a number of different ways. The architect can select systems and materials to achieve the look and texture he desires.

This chapter presents information on the most important materials and systems that are used in masonry veneer. Manufacturers of each product should be consulted for more detailed information on specific products and installation procedures associated with their products.

2.2 MASONRY UNITS

There are a variety of different veneer materials available for the building facade. Four general masonry categories are described in this chapter.

2.2.1 CLAY MASONRY UNITS

Numerous sizes, types, colors and textures of clay brick units are available for use as veneer. Depending on the units selected, the brick should be specified to meet one of the following American Society for Testing and Materials (ASTM) standards:

ASTM C62 *Standard Specification for Building Brick (Solid Masonry Units Made From Clay or Shale).*

ASTM C216 *Standard Specification for Facing Brick (Solid Masonry Units Made From Clay or Shale).*

ASTM C652 *Standard Specification for Hollow Brick (Hollow Masonry Units Made From Clay or Shale).*

ASTM C1088 *Standard Specification for Thin Veneer Brick Units Made From Clay or Shale.*

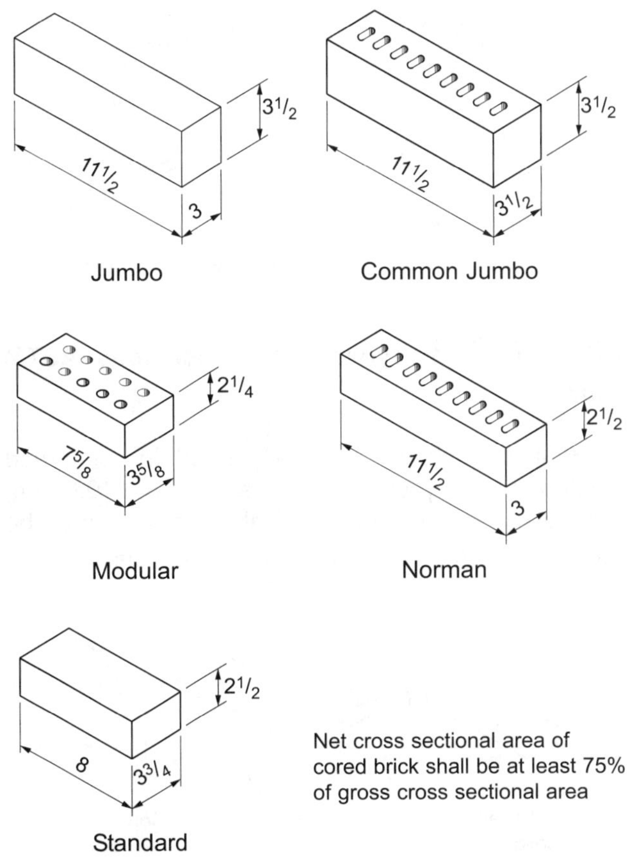

FIGURE 2.1 Typical face brick (ASTM C216).

Most anchored veneer installations use 3 to 4 in. (76 to 102 mm) wide facing brick as specified by ASTM C216. Building brick (ASTM C62) could also be used for anchored veneer but because of their appearance, color range, tolerances or chippage may not be as desirable.

Thin brick are manufactured and installed as adhered veneer construction and are typically between $7/16$ and 1 in. (11.1 and 25.4 mm) thick. Figure 2.2 illustrates typical thin brick units.

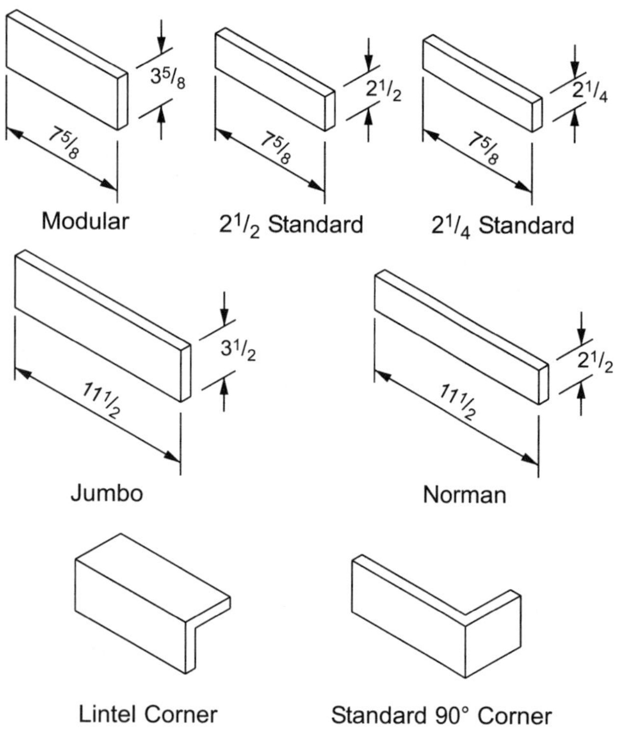

FIGURE 2.2 **Types of thin brick units (ASTM C1088).**

Hollow brick are generally used in reinforced grouted masonry construction and 4 in. (102 mm) nominal, $3\frac{1}{2}$ in. (89 mm) actual width units may be used in reinforced brick veneer installations.

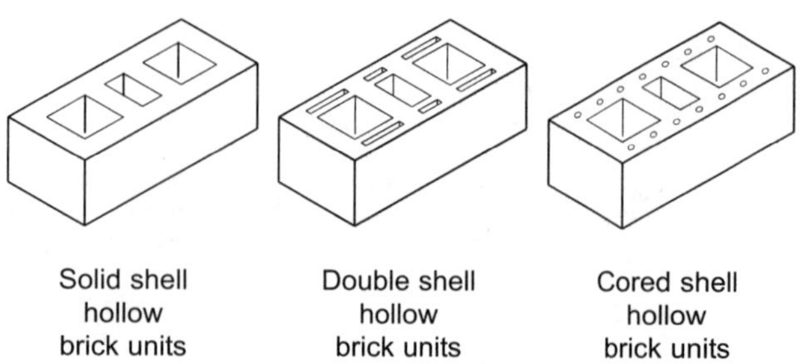

FIGURE 2.3 **Types of hollow (structural) brick (ASTM C652).**

2.2.1.1 GRADE OF BRICK

The three grades of brick, SW (Severe Weathering), MW (Moderate Weathering) and NW (Negligible Weathering), are classified according to their weathering resistance.

The effect of weathering on brick is related to the weathering index which is the product of the average annual number of freezing cycle days times the average winter rainfall in inches for any given locality. Grade requirements for exposures are listed in Table 2.1, and are described below. Figure 2.4 displays weathering indexes for the United States.

Grade SW (Severe Weathering) brick should be used where a high and uniform degree of resistance to both frost action and disintegration by weathering is desired and exposure is such that the brick may freeze in the presence of moisture.

TABLE 2.1 Grade Recommendations for Face Exposures[1]

Exposure	Weathering Index		
	Less than 50	50 to 500	500 and greater
In vertical surfaces:			
In contact with earth	MW	SW	SW
Not in contact with earth	MW	SW	SW
In other than vertical surfaces:			
In contact with earth	SW	SW	SW
Not in contact with earth	MW	SW	SW

[1]ASTM C62, Table 2

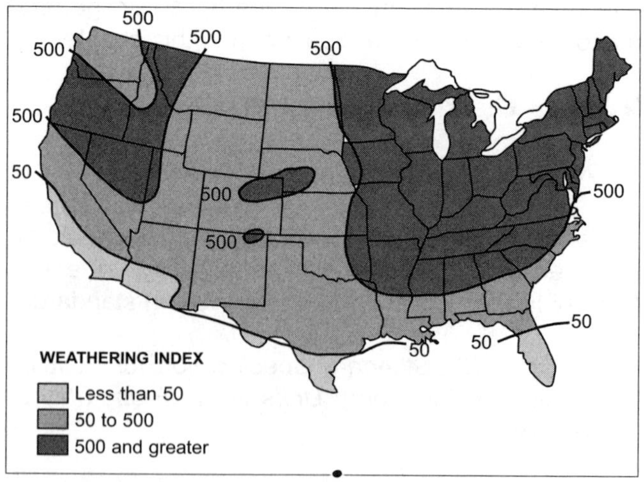

FIGURE 2.4 **Weathering index map of the United States.**

Grade MW (Moderate Weathering) brick should be used where moderate resistance to frost damage is acceptable and unsaturated freezing of the brick is permissible.

Grade NW (Negligible Weathering) brick is acceptable for interior masonry and backup for interior masonry. Grade NW units may disintegrate when subjected to freezing and thawing cycles.

Grade SW or grade MW may be substituted for grade NW brick, and grade SW brick may be substituted for grade MW brick. When the grade of brick is not specified, grade SW shall govern.

TABLE 2.2 Permissible Variations in Dimensions[1]

Specified dimension, in. (mm)	Maximum Permissible Variations from Specified Dimension, plus or minus, in. (mm)
Up to 3 (76), incl.	3/32 (2.4)
Over 3 to 4 (76 to 102), incl.	1/8 (3.2)
Over 4 to 6 (102 to 152), incl.	3/16 (4.8)
Over 6 to 8 (152 to 203), incl.	1/4 (6.4)
Over 8 to 12 (203 to 305), incl.	5/16 (7.9)
Over 12 to 16 (305 to 406), incl.	3/8 (9.5)

[1]ASTM C62, Table 3

Table 2.2 shows the allowable dimensional tolerances in the manufacturing of building brick. Minor indentations or surface cracks incidental to the manufacturing process, or minor chipping caused by the handling and delivery of brick shall not be grounds for rejecting the brick. Additionally, brick deliveries should contain not more than 5% broken brick. If a higher degree of precision from chips and cracks is required, the brick should be specified under ASTM C216, which contains guidelines of distance for visual inspection.

2.2.1.2 VISUAL INSPECTION

ASTM C216 states that all brick units shall be made of clay, shale or fireclay materials and that any admixtures or colors used in giving a special color to a unit shall be uniformly distributed through the entire body of brick. In other words, it may not be merely a surface color unless specifically agreed upon by the purchaser and seller. The standard also states that the brick shall be free of cracks and other defects that will interfere with the proper setting of the brick or that will impair the strength or the permanence of the construction.

Except for chips, FBX brick shall be free of cracks and other imperfections when viewed from a distance of 15 ft (4.6 m), and FBS or FBA brick shall not exhibit these imperfections when viewed from a distance of 20 ft (6.1 m).

This visual guideline applies to the installed materials, and can also be used for workmanship acceptance of the complete brick installation. Table

Table 2.3 Maximum Permissible Extent of Chippage from Edges and Corners of Finished Face or Faces onto the Surface[1]

Type	Percentage Allowed[2]	Chippage in in. (mm) in from		Percentage Allowed[2]	Chippage in in. (mm) in from	
		Edge	Corner		Edge	Corner
FBX	5% or less	1/8 – 1/4 (3.2 – 6.4)	1/4 – 3/8 (6.4 – 9.5)	95 to 100%	0 – 1/8 (0 – 3.2)	0 – 1/4 (0 – 6.4)
FBS[3] (Smooth)	10% or less	1/4 – 5/16 (6.4 – 7.9)	3/8 – 1/2 (9.5 – 12.7)	90 to 100%	0 – 1/4 (0 – 6.4)	0 – 3/8 (0 – 9.5)
FBS[4] (Rough)	15% or less	5/16 – 7/16 (7.9 – 11.1)	1/2 – 3/4 (12.7 – 19.1)	85 to 100%	0 – 5/16 (0 – 7.9)	0 – 1/2 (0 – 12.7)
FBA	to meet the designated sample or as specified by the purchaser, but not more restrictive than Type FBS (rough)					

[1] ASTM C216, Table 2

[2] Percentage of exposed brick allowed in the wall with chips measured the listed dimensions in from an edge or corner.

[3] Plain units are extruded brick with an unbroken natural die finish face.

[4] Textured units are extruded brick with the face sanded, combed, scratched, scarified, or broken by mechanical means such as wire-cutting or wire-brushing, or molded brick.

2.3 lists the ASTM criteria for acceptability of chippage in face brick.

2.2.2 CONCRETE MASONRY UNITS

Concrete units are increasingly popular as a veneer material due to the availability of split face, fluted, burnished and color tinted units.

Concrete brick or non-structural block veneer units can be specified under either of the following guidelines:

ASTM C55 *Standard Specification for Concrete Building Brick.*

ASTM C90 *Standard Specification for Loadbearing Concrete Masonry Units.*

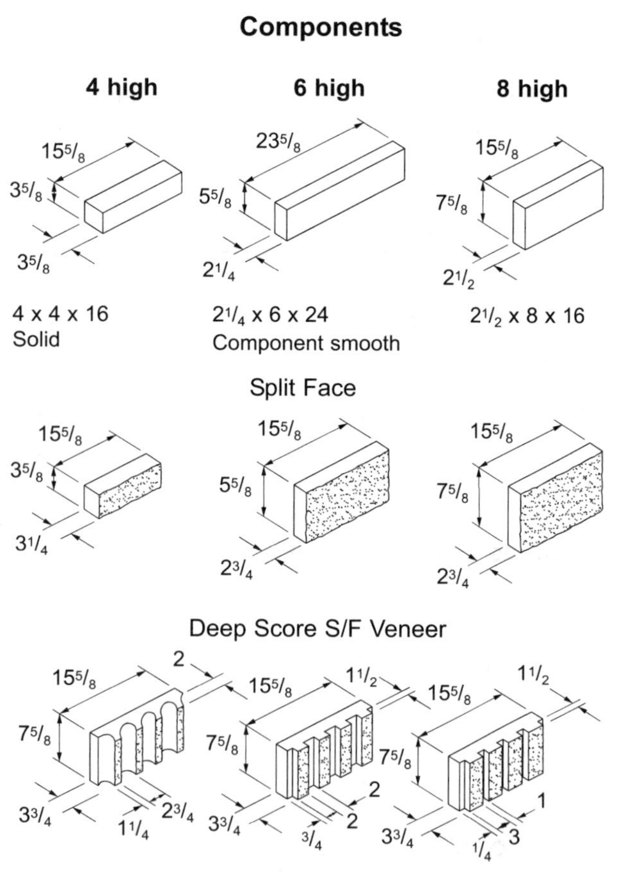

FIGURE 2.5 **Concrete masonry veneer units.**

In significant freeze-thaw locations, concrete brick should be specified as ASTM C55, which is comparable to a high strength concrete block.

ASTM C55 Grade N units offer high strength and resistance to moisture penetration in areas of severe frost action. The ASTM compressive strength requirements are 3500 psi (24.1 MPa) average for 3 concrete brick with the minimum compressive strength of 3000 psi (20.7 MPa) on any individual unit.

ASTM C55 Grade S units give moderate strength and resistance to moisture penetration and frost action. Strength requirements are 2500 psi (17.3 MPa) for the average of 3 concrete brick units, and the strength requirement on any individual unit must be at least 2000 psi (13.8 MPa).

Since the block manufacturer has discretion of what is considered a concrete brick, it is prudent to specify concrete veneer units under ASTM C90 where the freeze-thaw consideration is secondary. This avoids confusion since there is no absolute definition between concrete brick and solid concrete masonry units. At the same time the project is given a quality system that is more cost competitive in lieu of using high strength units that are not required.

Hollow concrete block are seldom used for veneer, but often used for back-up walls and are governed by:

ASTM C90 *Standard Specification for Loadbearing Concrete Masonry Units.*

2.2.2.1 COMPRESSION STRENGTH

ASTM C90, *Standard Specification for Loadbearing Concrete Masonry Units*, requires that the average compressive strength of 3 units, based on the net area of the unit, be at least 1,900 psi (13,100 kPa). In 1990, the Standard deleted the differentiation between Grade N and Grade S concrete block and required all block units under ASTM C90 designation meet the higher, old Grade N, Standard.

Grade N units were defined as acceptable for exterior use above or below grade, whereas Grade S units were limited to above grade exterior use provided the units are covered with a protective coating or not exposed to the weather.

For historical references, the widely used 1997 Uniform Building Code (UBC) Standard 21-4 was similar to ASTM C90, however Standard 21-4 did contain some intentional differences relating to the concrete masonry unit requirements. One of the differences is the recognition of both Grade N and Grade S by Standard 21-4.

Table 2.4 Strength and Absorption Requirements[1]				
Compressive Strength[2], min. psi		Water Absorption, Max. (Avg. of 3 units) with Oven-Dry Weight of Concrete, lb./ft[3]		
Average Net Area		Weight Classification		
Avg. of 3 Units	Individual Unit	Lightweight Less than 105	Medium Weight 105 to Less than 125	Normal Weight 125 or More
1900	1700	18	15	13

1 ASTM C90, Table 2.
2 Higher compressive strengths may be specified where required by design. Consult with local suppliers to determine availability of units of higher compressive strength.

Also, UBC Standard 21-4 specifically recognizes slumped block units as a structural unit. The slump textured block is widely used in the Southwestern United States to provide the traditional adobe appearance.

The amount of moisture a concrete unit absorbs can affect its strength. Table 2.4 lists the strength and absorption requirements for concrete masonry units.

The specified compression strength value that is common for ASTM C90 conforming units can be used to verify a design strength (f'_m) = 1,500 psi (10,342 kPa) for solid or partially grouted reinforced masonry walls. The definition of f'_m is the design value used in masonry. The actual compressive strength of the masonry system must not be less than the design value of f'_m. This value of 1,500 psi (10,342 kPa) is not the compression strength of the unit, it is the design value of the total masonry system.

Table 2105.2.2.1.2 of the 2006 International Building Code (2006 IBC), lists verified compression strengths of the masonry based on the compression strength of the concrete unit and the type of mortar used. The strength of the grout should conform to ASTM C 476 and the minimum compression strength of the grout must equal or exceed f'_m but not less than 2,000 psi (13.8 MPa) (2006 IBC Section 2105.2.2.1.2).

For example, when a masonry wall is designed for a compression strength of 1,500 psi (10,342 kPa), the concrete masonry units need to be 1,900 psi (13,100 kPa) using Type S or M mortar.

For high-rise load-bearing buildings, high cantilever retaining walls, tall slender walls with an *h/t* ratio between 30 to 50, or heavily loaded members, the strength of the concrete masonry units may require more than the basic minimum strength of ASTM C90. When high strength units are required, the units should be specified as follows:

"Concrete masonry units shall conform to the requirements of ASTM C90, except that the average unit net area strength shall be at least________psi and the specified compression strength of the masonry wall, solid grouted, shall not be less than the design strength, f'_m =________psi".

The percentage of desirable increase over the specified f'_m is dependent on the materials and experience in the area. Typically, the material strengths should be 25% to 33% greater than f'_m.

2.2.3 STONE

Hundreds of stone types and colors are available for use as veneer on buildings. In addition, the arrangement of the stone type and color should be considered. Figures 2.6 and 2.7 provide various patterns of rubble and ashlar stone.

Stone may be trimmed and cut to form regular patterns such as range, random or broken ashlar.

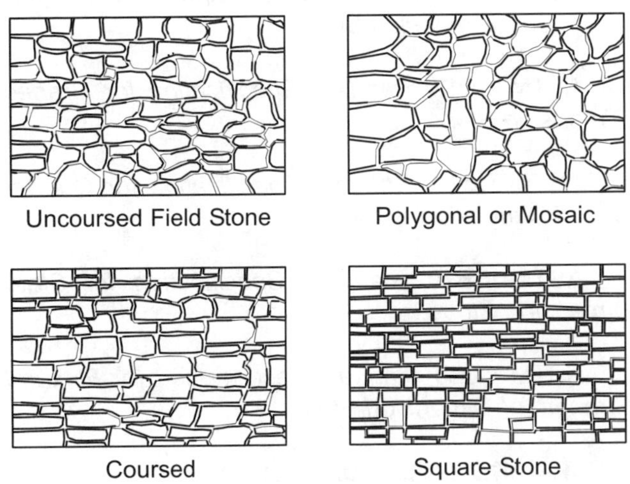

FIGURE 2.6 Rubble stone.

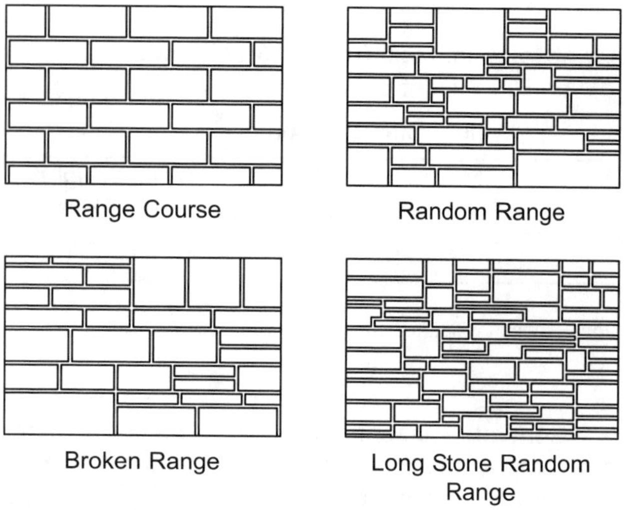

FIGURE 2.7 **Ashlar stone.**

2.2.3.1 ASTM STONE SPECIFICATIONS

In order to assure quality stone there are ASTM Standards that may be used and referred to:

ASTM C97 *Standard Test Method for Absorption and Bulk Specific Gravity of Dimension Stone.*

ASTM C99 *Standard Test Method for Modulus of Rupture of Dimension Stone.*

ASTM C170 *Standard Test Method for Compressive Strength of Dimension Stone.*

ASTM C503 *Standard Specification for Marble Dimension Stone (Exterior).*

ASTM C568 *Standard Specification for Limestone Dimension Stone.*

ASTM C615 *Standard Specification for Granite Dimension Stone.*

ASTM C616 *Standard Specification for Quartz-Based Dimension Stone.*

ASTM C629 *Standard Specification for Slate Dimension Stone.*

ASTM C880 *Standard Test Method for Flexural Strength of Dimension Stone.*

2.2.3.2 CLASSIFICATION OF STONE

Natural stone, such as marble and granite, is one of the most beautiful of earth's natural materials used in construction. The variegated surface of marble and the visual strength of granite have made these stones ideal for the creative designer.

2.2.3.2.1 GEOLOGICAL CLASSIFICATION

In the most general sense, masonry stone are rocks that form the earth's crust and fall into three genetic groups: *Igneous; Sedimentary; and Metamorphic.*

Igneous – When the earth formed, the crust was at one stage a viscous liquid skin that slowly cooled and hardened into igneous rock. Granite is a multi-colored intrusive igneous rock ranging in color from white to black. Granite is composed of quartz, feldspar, mica and ferro magnesium minerals. As igneous rock weathers, it deteriorates into deposited soils that then consolidated due to heat, pressure and cementitious action to produce sedimentary rock.

Sedimentary – Sedimentary rocks are formed by the process of cementing, consolidating crystallization and hardening of chemical solutions and biological deposits.

Metamorphic – When sedimentary rocks are subjected to increasing heat, pressure and shear, the minerals are then altered chemically and distorted or physically realigned to produce metamorphic rocks.

Table 2.5 lists some of the common masonry stone types and their geologic category, along with a description of the available finishes.

Metamorphic rocks may resemble their original sedimentary ancestors but are usually more crystalline and dense. Igneous rocks may also be metamorphosed by heat, pressure and shear but the changes are usually less drastic.

This manner of heat and pressure forms dense metamorphic rocks, differing widely from the igneous or sedimentary types. When limestones or dolomites crystallize, they form a metamorphic rock commonly called marble. All calcareous (calcium based) rocks, as well as some dolomitic and serpentine rocks that are able to be polished, are commercially called marbles.

Natural patterns within the stone can be highly varied and provide special features that make building stone a unique material. In addition to the pattern, the texture is varied, ranging from coarse fragments to the fine grains and crystalline structures. Texture also varies with the hardness of minerals composing the stone.

Table 2.5 Building Stone Surface Finish

Geological Category	Common Name	Finishes
1. Sedimentary	Sandstone Limestone Dolomite	A) Smooth (machine finished by saw, grinder or planer) B) Machine Tooled (with uniform grooves) C) Chat Sawn (non–uniform) D) Shot Sawn (irregular and uneven markings) E) Split Face (concave–convex) F) Rock Face (convex)
2. Metamorphic	Marble Serpentine Onyx Slate[1] Quartzite[1] Gneiss[2] Travertine[4]	A) Sanded B) Honed C) Polished D) Wheelabraded E) Bush–Hammered F) Split Face G) Rock Face
3. Igneous	Granite Syenite Diorite[3] Gabbro Andesite Basalt	A) Sawn B) Honed C) Polished D) Machine Tooled (4–cut, 6–cut, chiseled, axed, pointed, etc.) E) Flamed F) Sand Finished G) Split Face H) Rock Face

1 Slate and Quartzite cannot be polished.
2 Gneiss will take all of the finishes of marble and may also be flame finished.
3 Diorite will not take flame finish.
4 Travertine is actuallly a limestone but is classified with marbles for surface finishes. Travertine finishes include filled, partially filled and unfilled.

Pattern, texture, and color are all affected by how the stone is fabricated and finished. Granites hold color and pattern, while limestone color changes with exposure. Textures may range from rough and flamed finishes to honed or polished surfaces. As a 'hard' stone, granite has the ability to become a highly polished finish which can be maintained over the life of the stone.

2.2.3.2.2 TEXTURE OF QUARRIED STONE

The term "texture", as applied to marble, relates to the size, shape, degree of uniformity and arrangement of the component grains or crystals. The texture or grain pattern can be:

Equigranular – grains of approximately the same size, such as limestone.

Inequigranular – grains of markedly unequal sizes such as granite.

Porphyritic – relatively large, coarse crystal called phenocrysts of one or more mineral components in a ground mass of markedly finer texture such as granite.

Interlocking – where grains with irregular boundaries interlock by mutual penetration such as granite and breccia.

Mosaic (or Granulitic) – closely packed grains with smooth to moderately irregular, non-interlocking mutual boundaries such as feldspar and pyroxene.

Elastic – naturally cemented fragmental grains but without interlocking or mosaic relation such as quartzite.

Grano Balstic – a granular mosaic texture in which the grains are tightly compacted, the minerals are dominantly of equidimensional kinds and present irregular mutual boundaries such as granite.

Table 2.6 lists some common stone veneer used in the United States.

2.2.3.2.3 CLASSIFICATION OF FABRICATED STONE FOR SOUNDNESS

Building stones are classified into different groupings based on their geological makeup, chemistry and appearance. Most stones suitable for use as slab veneer are classified as granite, marble, travertine (sometimes called a marble), limestone and sandstone.

Granite is described by ASTM as a visibly granular rock generally ranging in color from pink to light or dark gray and consisting mostly of quartz and feldspars accompanied by one or more dark minerals. Granite is generally a very sound stone with consistently high working qualities and relatively uniform color variations.

ASTM describes **Limestone** as a rock of sedimentary origin composed principally of calcium carbonate, or the double combination of carbonate of calcium and magnesium, or a combination of both. Limestone that is capable of taking a polish is often categorized as commercial marble and sold as either limestone or marble. **Travertine** is a variety of limestone which is sometimes categorized as commercial marble. Travertine is a product of chemical precipitation from hot spring which results in a cellular stone usually concentrated in thin layers that often display a stalactic structure.

Marble is described by ASTM as stone capable of taking a polish. Stones in this group comprise a variety of textures. Most have an interlocking texture and a wide range of grain sizes. Calcite and dolomite marble are formed by the heat and pressure of metamorphism, and usually acquire a distinctive crystalline texture. Mineralogically they are comprised mainly of calcite and dolomite, either singly or in combination. Some compact dense **limestones** that can take a polish are commercially classified as marble.

Sandstone is sedimentary rock composed mostly of mineral and rock fragments that are "sand"

Table 2.6 West Coast Veneer Stone

Name	Type	Color	Texture	Coverage (sq. ft./ton)
Palos Verdes	Sedimentary Stone	Off White to Gray	Flat to Uneven	40–45
Bouquet Canyon	Granite Schist	Tan to Rust	Flat to Uneven	40–45
Santa Maria	Sedimentary Limestone	Cream to Rust	Flat	40–45
Drift Stone	Wallostonite	Brown to Black	Rough to Rugged	40–45
Black Lava	Lava	Black	Rough to Rugged	45–50
White Marble	Limestone	White	Irregular	40–45
Arizona Cut Wall	Sandstone	Tan and Red	Smooth	45–50
Featherock	Pumice	Gray–Black	Rough	150–180
Slate	Sedimentary Shale	Gray–Green–Plum	Smooth	300–350
Texas Shell	Oolitic Limestone	Cream	Irregular	45–50
Texas Lime	Oolitic Limestone	Cream	Smooth	45–50
Whitewater Canyon	Unknown	Red w/Yellow	Irregular	45–50
Mariposite	Quartzite	Green w/White	Irregular	25–30
Grimes Canyon	Unknown	Red w/Yellow	Irregular	45–50
Desert Bark	Shale	Red to Brown	Irregular	45–50
Santa Rita	Sedimentary Limestone	Cream to Gray	Flat	45–50
Apache Stone	Shale	Green–Red–Pink	Flat	45–50

sized 0.06 to 2.0 mm, and having a minimum of 60 percent free silica. Sandstone must have a minimum compressive strength of 4,000 psi (27.6 MPa). The variety of colors is due to the mixture of minerals. The character of cementing matter more than the grains determine the appearance and durability of the stone.

The selection of filling materials should be left to the finisher, who typically knows what materials work best with these stones. Common materials used are natural gray or tinted Portland cement and clear or colored epoxy or polyester resins. If epoxies are used, interior or exterior grade epoxy should be specified. The choice of filler material should be compatible with the environment.

2.2.4 TERRA COTTA

Terra cotta is a Latin word that means "baked earth". This ancient process has been used for thousands of years. A special aged clay is used which can be molded into simple building units or into very ornate three–dimensional figures. This clay is fired at high temperatures providing a compact and vitrous (glassy) surface texture.

As a ceramic tile, terra cotta should be specified by the following:

ASTM C126, *Standard Specification for Ceramic Glazed Structural Clay Facing Tile, Facing Brick, and Solid Masonry Units.*

2.2.4.1 CHARACTERISTICS

Today terra cotta is used as a ceramic veneer; it is fired clay, fired to vitrification and impervious to moisture. The clays are a composition of fire clays, ball clay, feldspars, sand and grog. They are processed and ground to enable a smooth surface. Both hand pressed and extruded shapes are manufactured.

Typically terra cotta units are formed in hollow blocks, open at the back with several web stiffeners to reduce the weight of terra cotta manufactured for cladding. They can be ashlar units or have more intricate profiles. Typically, the units are glazed on surfaces that will be exposed when the unit is set in the wall.

With the addition of ceramic stains, terra cotta can be made in a variety of different colors, although a brownish orange is the most common terra cotta body color.

Terra cotta is of ancient origin, but it is also used extensively today and manufactured under controlled high quality conditions. The body of the clay unit is fired at a temperature of 2100°F (1150°C) to fuse the clay body and the glaze finish. Reds, yellows and orange glazes are fired at somewhat lower temperatures, 1650°F to 1950°F (900°C to 1065°C).

Gold, silver and platimums are fired at still lower temperatures over the matured glazes at 1350°F to 1600°F (730°C to 870°C).

Terra cotta comes in an infinite variety of shapes and sizes, and most units are anchored.

Adhered terra cotta are thin masonry units, up to $1^1/_4$ in. (31.8 mm) in thickness, including back ribs. The back ribs are necessary in order to provide a secure bond with mortar.

Adhered terra cotta should never be applied to structural masonry since the masonry units will expand and contract during normal moisture retention and drying cycles. A waterproof membrane should first be applied to the concrete masonry surface, then an expanded metal lath wire mesh and a scratch and brown coat of plaster applied prior to the adhered installation of terra cotta.

In the past, terra cotta units were laid up in the wall with narrow mortar joints using a cement–lime mortar. In some buildings, the hollow backs of the units were filled with masonry and mortar, with the fill keyed into the masonry backup wall. Also, metal ties were used providing positive anchorage of the terra cotta.

FIGURE 2.7 **Terra cotta cornice.**

Today, units are used as cladding and are supported by steel shelf angles at floor levels, above wall openings and at projections such as cornices. Steel straps are set into slots at the back of each unit for horizontal support and tied to the anchorage system of the supporting wall. The projecting cornice units are suspended from the anchorage system by steel bolts, which hold horizontal bars threaded through holes formed in the webs of the terra cotta block. Well engineered systems will withstand earthquakes and severe weather conditions.

2.3 MORTAR

Mortar is a basic component of masonry. Some claim that mortar holds the units apart while others claim it holds the masonry units together. It actually does both. Mortar is specified by:

> ASTM C270 *Standard Specification for Mortar for Unit Masonry.*

Depending on the proportions of each of the constituent materials, the mortar properties will change accordingly. Mortars rich in cement (i.e. Type M mortar) will be strong in compression. By contrast, a mortar with a high percentage of lime (i.e. Type O mortar) will be workable since high lime mortars exhibit greater plasticity than low lime mortars. The relative compressive strengths of Types N and O mortar however will be reduced. The proper mortar must be selected upon the particular needs and conditions that are required.

Properties of bond strength, workability and moisture prevention should be considered when specifying the type of mortar for a project. Portland cement-lime mortar may be preferred over masonry cement mortars since its performance is usually better and more predictable. Air-entraining agents and other ingredients that prevent good bond properties are often contained in masonry cements.

Ideally the mortar should be weaker than the units so that if shrinkage or temperature stresses occur, any potential cracks will form in the mortar joints and not the units.

These cracks in the mortar joints can be repaired by pointing while cracks in the units may require removal and replacement of the units.

Table 2.7 Mortar Proportions for the Unit Masonry[1]

Mortar	Type	Proportions by Volume (cementitious materials)								Aggregate Ratio Measured in a Damp Loose Condition
		Portland Cement or Blended Cement	Masonry Cement M	Masonry Cement S	Masonry Cement N	Mortar Cement M	Mortar Cement S	Mortar Cement N	Hydrated Lime or Lime Putty	
Cement–lime	M	1	–	–	–	–	–	–	$\frac{1}{4}$	Not less than $2\frac{1}{4}$ and not more than 3 times the sum of the separate volumes of cementitious materials
	S	1	–	–	–	–	–	–	over $\frac{1}{4}$ to $\frac{1}{2}$	
	N	1	–	–	–	–	–	–	over $\frac{1}{2}$ to $1\frac{1}{4}$	
	O	1	–	–	–	–	–	–	over $1\frac{1}{4}$ to $2\frac{1}{2}$	
Mortar cement	M	1	–	–	–	–	–	1	–	
	M	–	–	–	–	1	–	–	–	
	S	$\frac{1}{2}$	–	–	–	–	–	1	–	
	S	–	–	–	–	–	1	–	–	
	N	–	–	–	–	–	–	1	–	
	O	–	–	–	–	–	–	1	–	
Masonry cement	M	1	–	–	1	–	–	–	–	
	M	–	1	–	–	–	–	–	–	
	S	$\frac{1}{2}$	–	–	1	–	–	–	–	
	S	–	–	1	–	–	–	–	–	
	N	–	–	–	1	–	–	–	–	
	O	–	–	–	1	–	–	–	–	

[1]2006 IBC Table 2103.8(1)

2.3.1 PROPORTION SPECIFICATION

Proportion specifications limit the amount of the constituent parts by volume. Water content, however, may be adjusted by the mason to provide proper workability under various field conditions. When the proportions of ingredients are not specified, the proportions by mortar type must be used as given in Table 2.7.

Mortar other than those approved in Table 2.7 may be used when laboratory or field tests demonstrate that the mortar, when combined with the masonry units, will achieve the specified compressive strength of the masonry assemblage, f'_m.

The most common portland cement–lime mortar proportions by volume are in Table 2.8.

Table 2.8 Common Portland Cement-Lime Mortar Mixes

Type M Mortar:	1 part portland cement $^1/_4$ part lime $3^1/_2$ parts sand
Type S Mortar:	1 part portland cement $^1/_2$ part lime $4^1/_2$ parts sand
Type N Mortar:	1 part portland cement 1 part lime 6 parts sand
Type O Mortar:	1 part portland cement 2 parts lime 9 parts sand

2.3.2 MORTAR MATERIALS

The principal mortar ingredients are cement, lime, sand and water, each making a unique contribution to a mortar's performance. Cement contributes durability, high-early strength and high compressive strength to mortar. Lime contributes to workability, water retentivity and elasticity. Both contribute to bond strength. Sand acts as a filler and contributes to the strength. Water is the ingredient which creates a plastic, workable mortar and is required for the hydration of the cement.

2.3.2.1 CEMENTS

Three types of cement are permitted to be used in mortar; portland cement, masonry cement and mortar cement. Mortar cement has been developed as ASTM C1329, *Standard Specification for Mortar Cement* and may be used in high seismic applications.

2.3.2.1.1 PORTLAND CEMENT

The basic cementitious ingredient in mortar is portland cement. This material must meet the requirements of ASTM C150, *Standard Specification for Portland Cement*. In mortar, the type of portland cement is limited to Type I, II or III. The use of air–entraining portland cement (Type IA, IIA or IIIA) is not recommended for masonry mortar because air entrainment can reduce the bond between mortar and the masonry units.

Portland cement is the primary adhesive material and based on the water/cement ratio can produce high strength mortars. Hydrated lime is used in conjunction with portland cement to provide the desired strength, bond, workability and board life (board life is defined as the time during which mortar is still plastic and workable).

2.3.2.1.2 MASONRY CEMENTS

Masonry cement is a proprietary blend of portland cement and plasticizers such as ground inert fillers and other additives for workability. Masonry cement must meet the requirements of ASTM C91, *Standard Specification for Masonry Cement* and is available in Types M, S and N.

The use of masonry cement for mortar is not permitted in lateral load-resisting masonry in Seismic Design Categories D, E and F.

Masonry cements can be proprietary with specific ingredients not disclosed, therefore, the MSJC Code imposes the seismic restriction.

2.3.2.1.3 MORTAR CEMENTS

Mortar cement is also a portland cement based material which meets the requirements of ASTM C1329, *Standard Specification for Mortar Cement*. Mortar cement may be used for mortar in all Seismic Design Categories.

Unlike masonry cement, mortar cement was developed specifically for use in seismic application, therefore, the lateral load–restriction does not apply.

Manufacturers of this material were deliberate in developing a cement that would satisfy the requirements of structural engineers and building officials.

2.3.2.2 HYDRATED LIME

Hydrated lime is manufactured from calcining limestone (calcium carbonate with the water of crystallization, $CaCO_3H_2O$). The high heat generated in the kiln drives off the water of crystallization, H_2O, and the carbon dioxide, CO_2 resulting in quicklime, CaO.

The quicklime can then be slaked by placing it in water thus making hydrated lime, lime putty or slaked lime $Ca(OH)_2$. The hydrated lime can then be dried and ground, producing a white pulverized hydrated lime which is sacked and used in mortar.

Hydrated lime can be used without delay, making it more convenient to use than quicklime.

Hydrated lime must conform to ASTM C207, *Standard Specification for Hydrated Lime for Masonry Purposes*, and is available in Types S, SA, N and NA. Types S and N hydrated limes contain no air entraining admixtures. Types NA and SA limes provide more entrained air in the mortar than allowed by ASTM and therefore may not be used. Additionally, unhydrated oxides are not controlled in Type N or NA limes thus making only Type S hydrated lime suitable for masonry mortar.

Lime in mortar provides cementitious properties and is not considered to be an admixture. Used in mortar it:

1. Improves plasticity or workability of the mortar.

2. Improves the water tightness of the wall.

3. Improves the water retentivity or board life of the mortar.

2.3.2.3 MORTAR SAND

For masonry mortar prepared by the proportion method, sand aggregate must conform to ASTM C144, *Standard Specification for Aggregate for Masonry Mortar*.

Sand used in preparing mortar can be natural or manufactured. Manufactured sand is obtained by crushing stone, gravel or air–cooled blast–furnace slag. It is characterized by sharp and angular particles producing mortars with workability properties different than mortars made with natural sand which generally have round, smooth particles.

Sand gradation is most often specified or defined by referring to a standard sieve analysis. For mortar, sand is graded within the limits given in Table 2.9.

Table 2.9 Sand for Masonry Mortar[1]

Sieve Size	Percent Passing	
	Natural Sand	Manufactured Sand
No. 4	100	100
No. 8	95 to 100	95 to 100
No. 16	70 to 100	70 to 100
No. 30	40 to 75	40 to 75
No. 50	10 to 35	20 to 40
No. 100	2 to 15	10 to 25
No. 200	0 to 5	0 to 10

[1] Based on ASTM C144, Section 4.

Sand should be free of harmful amounts of deleterious substances and organic impurities. ASTM C144 provides guidelines on determining if an aggregate has excessive impurities.

Concrete sand should not be used in mortar since the maximum grain size is too large. Additionally, the fine particles which are needed in masonry sand have often been washed out of concrete sand thus creating harsh, coarse sand unsuitable for mortar. Mortar sand needs at least 5% fines which pass the No. 200 sieve to aid plasticity, workability and water retention of mortar.

Mortar sand, like all mortar ingredients, should be stored in a level, dry, clean place. Ideally, it should be located near the mixer so it can be measured and added with minimum handling and can be kept from contamination by harmful substances.

2.3.2.4 WATER

Water must be clean and free of deleterious amounts of acids, alkalies or organic materials. Water containing soluble salts such as potassium and sodium sulfates should be avoided since these salts can contribute to efflorescence.

2.3.2.5 ADMIXTURES

There are numerous admixtures which may be added to mortar to affect its properties. One of these, called a retarding set admixture, delays the set and stiffening of mortar. The set may be delayed for 36 hours or more if desired.

There are also admixtures used to replace lime. These may be air entraining chemical or pulverized fire clay or bentonite clay to provide workability. Care should be taken with these admixtures since the bond between the mortar and the masonry units may be affected.

The use of any admixture must be acceptable to the architect or engineer and should also be acceptable to the building official.

2.3.2.6 COLOR

Mortar colors are normally mineral oxides or carbon black. Iron oxide is used for red, yellow, and brown colors; chromium oxide for green, and cobalt oxide for blue colors. Commercially prepared colors for mortars also offer a wide variety of colors and shades.

The amount of color additive depends on the color and intensity desired. Typically, the amount of color additive ranges from 0.5% to 7.0% for the mineral oxides with a maximum of 2% for carbon black. These percentages are based on the weight of cement content and the maximum percentages are far greater than the amounts of color additives usually required.

MSJC Specification limits the amount of color, based on percentage by weight of cement, to 10% mineral oxide pigment and 2% carbon black pigment for portland cement–lime mortars and 5% mineral oxide pigment and 1% carbon black pigment for masonry cement and mortar cement mortar.

Mixing time of the mortar should be long enough for a uniform, even color to be obtained and should be the same length of time for every mortar batch. Additionally the mixing sequence should be the same for each batch. MSJC Specification requires that mortar be mixed between 3 to 5 minutes.

Retempering of colored mortar should be kept to a minimum to reduce the variations in color of the mortar. For the best results, colored mortar should not be retempered.

Finally, the source, manufacturer and amount of each ingredient should remain the same for all colored mortar on a project in an effort to obtain uniform color throughout. Prepacked mineral color additives that can be added to the mix based on full sacks of portland cement will minimize mortar color variation.

2.3.3 PRE-BLENDED MORTAR

Mortar can also be factory pre-blended and stored at the jobsite in sacks or silos. Some silo systems introduce water to the dry mortar mix in an auger screw at the base of the silo, while other silo systems discharge the dry mortar mix directly into a conventional mixer.

Pre-blended dry mortar is also available in sacks, which may be beneficial in keeping project debris at a minimum. This packaging method can be especially useful in limited working areas, such as parking garages.

When factory blended mortar is used, manufacturers certification of the type of mortar is recommended.

2.4 TYPES OF MORTAR JOINTS

Figure 2.8 provides basic terms for mortar joints. The finished bed and head joints are cut, troweled, or tooled.

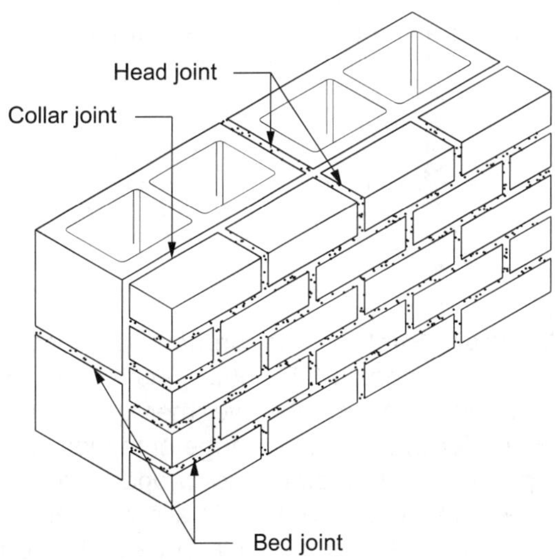

FIGURE 2.8 **Basic terms for mortar joints.**

Shown in Figure 2.9 are nine examples of commonly used mortar joints. Each joint provides a different architectural appearance to the wall. Since some joints provide poor weather resistance, care must be taken in the selection of the proper type of mortar joint. Joints with ledges such as weather, squeezed, raked, and struck joints perform poorly in exterior applications and may allow moisture penetration. Tooled joints are recommended for exterior applications since tooling compacts the mortar tightly preventing moisture penetration.

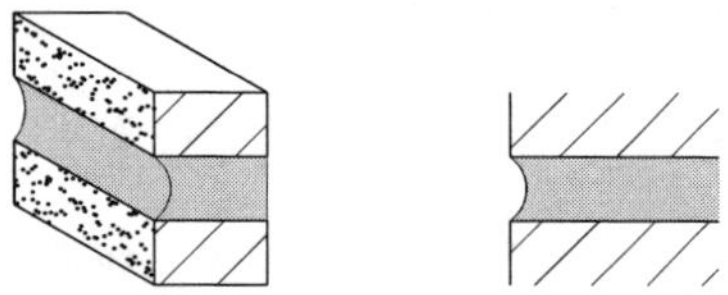

a. **Concave Joint** – It is the most common joint used. The tooling works mortar tight into the joint, compressing the mortar, producing a weather joint. The joint emphasizes the masonry unit pattern and conceals small irregularities in laying the unit.

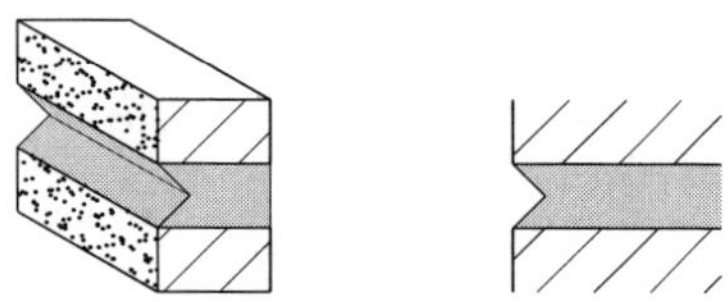

b. **"V" Joint** – Tooling works the mortar tight and provides a weather joint. However the "notch" of the "V" can be a point of discontinuity and cracks may develop which allow water migration. This joint emphasizes the masonry unit pattern and conceals small irregularities in laying, while providing a line in center of mortar joint.

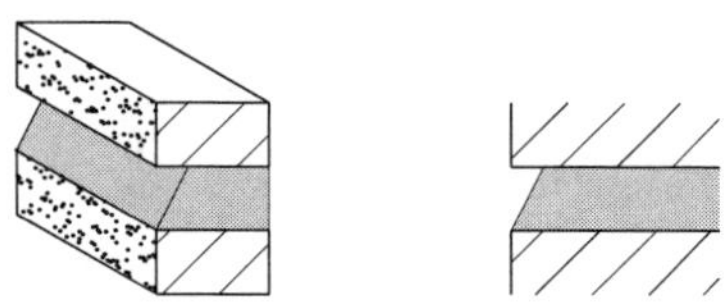

c. **Weather Joint** – The primary purpose is to emphasize horizontal joints. This type of joint is a marginally acceptable weather–type joint. The reason for this is the top ledge of the joint acts as drip ledge. If the joint is not properly tooled, the surface tension of water will allow water to pool at the drip ledge and the water can migrate back into the mortar.

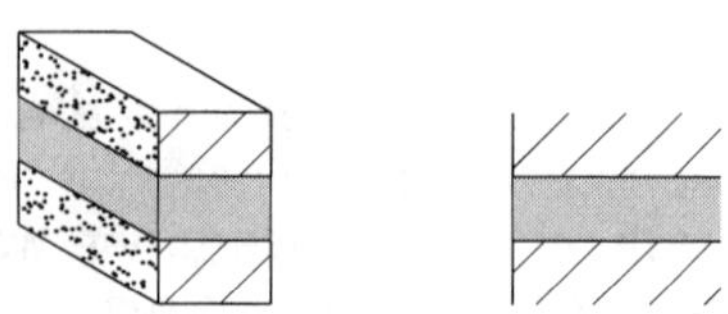

d. **Flush Joint** – This joint is used where the wall is to be plastered. Special care is required to make the joint weatherproof. Mortar joint must be compressed to assure intimate contact with the masonry unit. Not recommended for exposed exterior use.

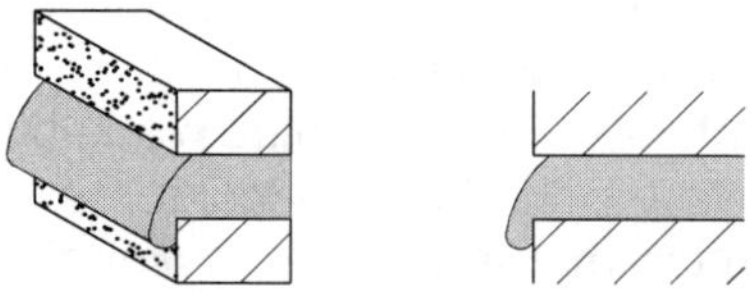

e. **Squeezed Joint** – This type of joint provides for a rustic, high texture appearance. Satisfactory for interior use and exterior fences. Not recommended for exterior building walls, since no weather resistance is created because the mortar is not compressed back into the joint. Also the top ledge allows for pooling of the water.

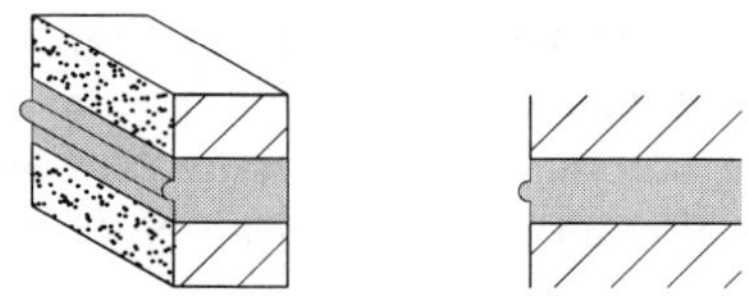

f. **Beaded Joint** – This is a special effect, poor exterior weather joint due to exposed ledge and not recommended.

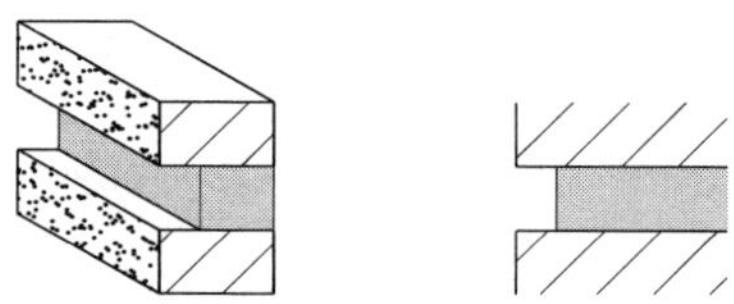

g. **Raked Joint** – This joint type strongly emphasizes the units. Poor weather joint and not recommended if exposed to weather unless tooled at bottom of mortar joint. Pooling of water can occur at the top ledge (surface tension properties of water) and the bottom ledge.

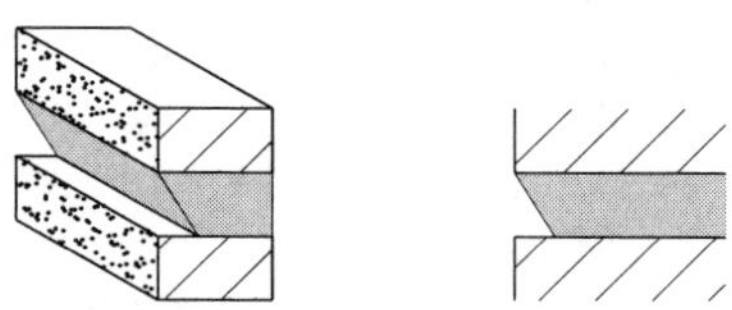

h. **Struck Joint** – This joint type is used to emphasize horizontal lines. Poor weather joint, therefore not recommended as water will penetrate on lower ledge.

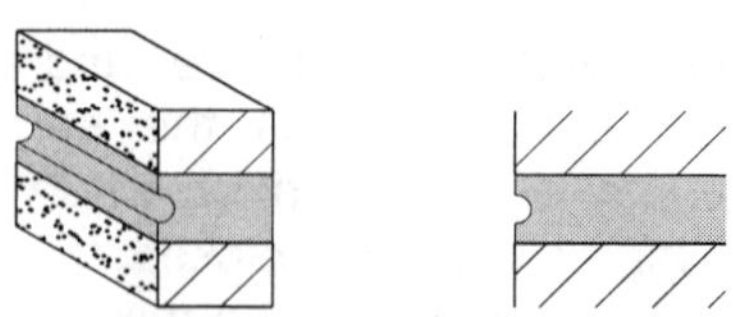

i. **Grapevine Joint** – This joint shows a horizontal indentation. Same limitations as flush joint.

FIGURE 2.9 Mortar type joints.

2.5 STRUCTURAL SUPPORT BACKUP MATERIALS

The structural support of a veneer system should have the same or very similar material properties as the veneer. The reason is compatibility. If the veneer and the backup system respond in similar manners to changes in the environment, then the tendency for cracking, movement differences, and the build up of secondary stresses are minimized. One of the best and obvious ways to assure the compatibility of material is to build the backup system of the same material as the veneer.

2.5.1 STRUCTURAL MASONRY BACKUP

An easy and rigid method to support veneer units is with a structural masonry backup. If the veneer is brick, then the backup can be made of the same materials and there is little concern about dissimilar material properties.

A rigid system deflects little. Since the backup system deflects very little and the veneer is attached to the backup system, then veneer will have limited deflection. The limited deflection of the veneer reduces the possibility of water intrusion through flexural cracks in the veneer.

Traditionally, structural masonry backup systems have been constructed with Concrete Masonry Units.

Clay brick, or hollow clay brick, can also be used to support veneer. This is seldom considered due to the increased cost. Instead, two wythes of brick with one wythe being the veneer brick facing have been used.

Anchored veneer can easily be attached to structural masonry with a variety of wall ties or joint reinforcement. A prime advantage of this type of system is the ease in which insulation can be placed in the cavity between the wythes of masonry. Moreover, when properly flashed and drained with weep holes, the system can prevent water intrusion into the building.

Prior to placing the anchored veneer the structural masonry backup must be waterproofed. Frequently, this is done by painting the backup with bituminous tar or by covering with 30 pound (13.6 kg) felt. This added measure helps mitigate the moisture penetration through the masonry.

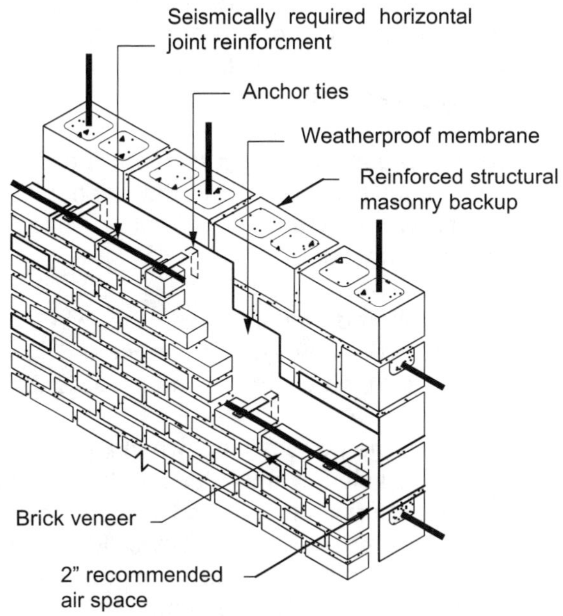

FIGURE 2.10 **Structural Masonry backup system.**

When clay brick veneer is installed with a concrete masonry backup, a provision to allow for differential movement between the materials must be included.

Clay masonry expands with moisture and temperature over a long period of time while concrete masonry shrinks and creeps. Accordingly, solidly bonding clay masonry to concrete masonry is not recommended except for small installations.

A space or cavity should exist between the veneer and the concrete masonry units. Veneer ties, which allow for minor differential movement between the materials, are required. Expansion, shrinkage and isolation joints should also be carefully located so that each material can move relatively free of the other.

2.5.2 STRUCTURAL CONCRETE BACKUP

Poured and precast reinforced concrete structures are often veneered with masonry. A few items require special attention to avoid problems with concrete backup walls:

1. When the veneer is to be adhered to the concrete, remove all form release agents (such as oil, wax or grease) or curing agents that are present on the concrete. These substances act as bond breakers and can prevent the mortar from bonding to the

concrete. If any such substances were used, the surface of the concrete should be sandblasted or water blasted until all such materials are removed.

2. As with masonry backup systems, movement joints should be placed in the veneer at all locations where movement joints occur in the concrete backup.

3. For anchored veneer, cast the wire ties, anchor slots and other embedded items directly into the concrete when practical.

   A dovetail tie system is shown in Figure 2.11. A "dovetail slot" is embedded in the concrete and veneer ties are inserted into the dovetail slot. The dovetail slot allows for vertical adjustment during construction, as well as vertical expansion movement.

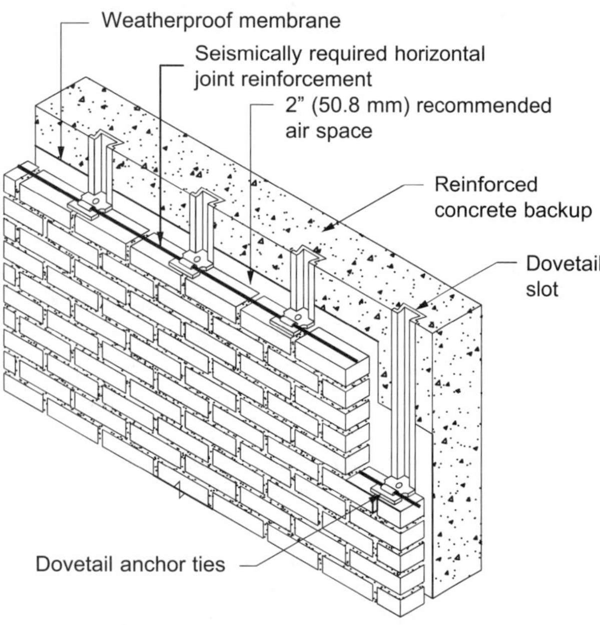

FIGURE 2.11 **Concrete backup system.**

2.5.3 WOOD STUD BACKUP

Wood studs are used to backup masonry veneer primarily on residential structures and low rise commercial buildings. Multistory wood framed buildings can be veneered but accommodations must be provided to allow for lateral load support, vertical shrinkage and movement of the wood structure.

For single story buildings, wood studs are typically located at 16 in. (406 mm) on center although spacing of 24 in. (610 mm) may be used when the loads are small or when larger studs are utilized. Deflections of the wood studs should be limited to no more than *h*/600 or *h*/720 to reduce cracking in the masonry veneer.

Only quality (grade 2 or better), seasoned lumber should be used. Wet and green lumber should not be used since shrinkage due to drying may damage not only the veneer but also interior drywall and other surface treatments. Excessively warped, bowed or damaged studs should also be rejected.

Design of the wood stud backup system must comply with either IBC Chapter 23 or the National Design Specification by the National Forest Products Association.

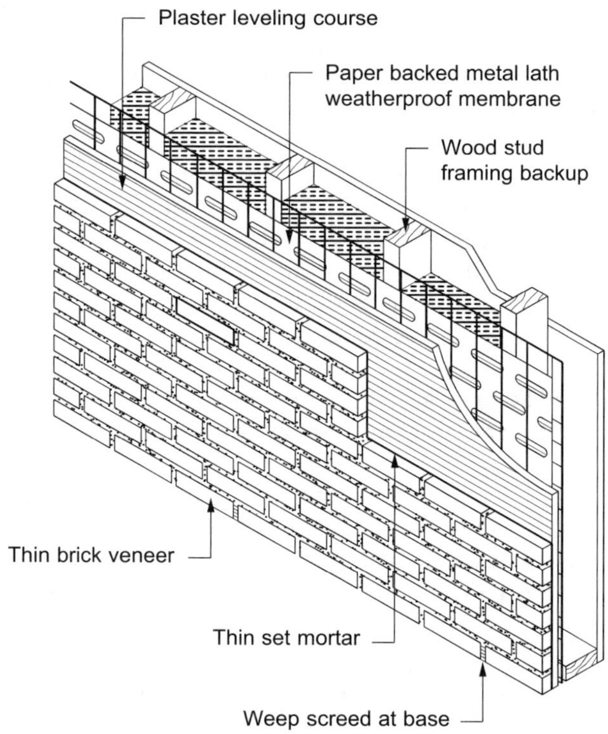

FIGURE 2.12 **Wood stud backup system for adhered veneer.**

2.5.4 STEEL STUD BACKUP

Steel studs have been used to support veneer construction for several decades. Initially, flexible and light steel studs were used and little attention was paid to properly waterproofing the veneer/steel stud system. As a result of these initial practices, design changes were quickly implemented and attention was given to design and construction quality control thereby yielding a reliable steel stud/veneer system.

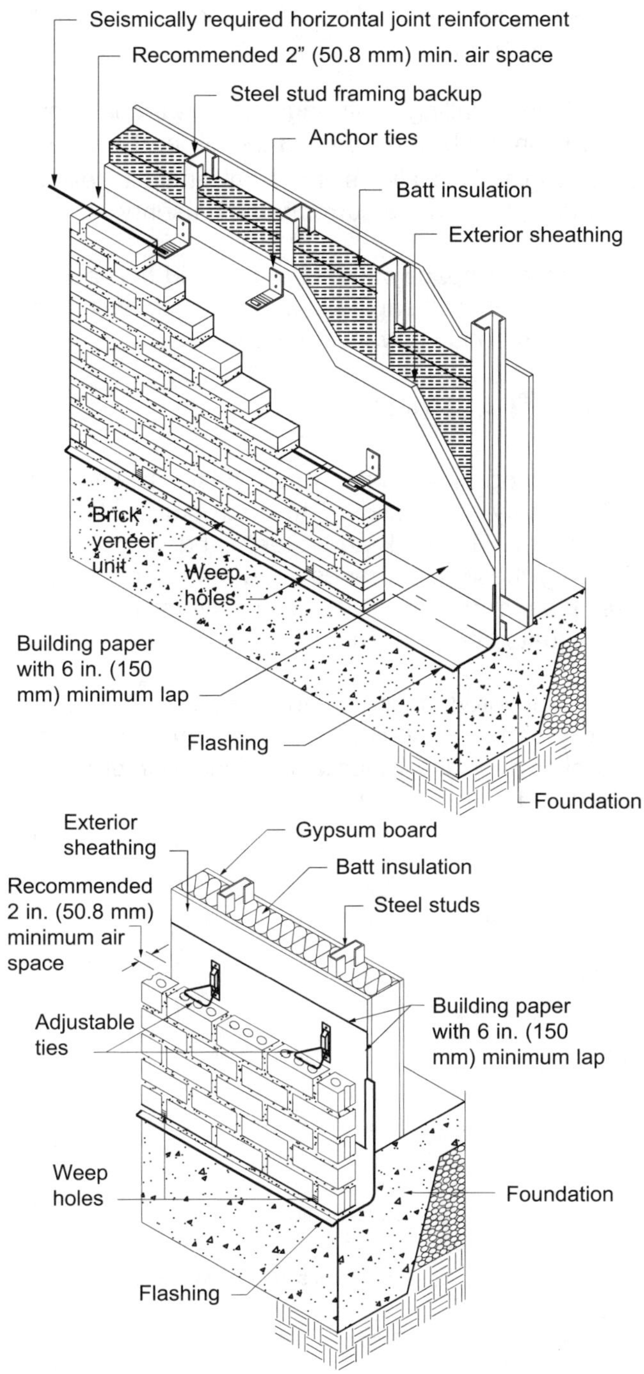

FIGURE 2.13 **Steel stud backup system.**

Recommendations from the Brick Industry Association, Technical Notes 28B, *Brick Veneer/Steel Stud Walls* and Western States Clay Products Association, *Design Guide for Anchored Brick Veneer Over Steel Studs* include:

1. The brick veneer/steel stud wall system should be detailed as a panel wall, or fully supported at each story height by the structural frame through shelf angles.
2. Limit the deflection of the steel studs to $h/600$ of the height of the stud backup when considered alone as a full lateral design load. This ensures that the studs will be sufficiently stiff so that the veneer will be adequately supported with only minor cracking from bending.
3. Use at least 18 gauge (1.2 mm) metal studs to allow for adequate screw thread grip. However, the deflection criteria may require a minimum of 16 gauge (1.5 mm) metal stud, except for lightly loaded and dry areas, such as building interiors where 18 gauge (1.2 mm) studs may be adequate.
4. Steel studs should be galvanized coated to conform to ASTM A525, Grade G-90.
5. Rigid sheathing should be securely fastened to both sides of studs. The sheathing must be properly detailed and attached.
6. Horizontal bracing at mid-height for added stiffness is recommended.

The design of the steel stud system should meet the requirements of either the IBC or the Metal Lath/Steel Framing Association design criteria.

2.6 SHELF ANGLES

Corrosion resistant shelf angles are provided to support only the weight of the veneer. Thus, a backup system is still required to support the veneer for lateral loads.

Shelf angles are also used to support the weight of the veneer above window and door openings.

Shelf angles are typically installed at each floor level of multistory buildings in high seismic exposure so that the vertical loads on the veneer are not excessive and so the structural frame can move slightly without damaging the masonry veneer. A $^{1}/_{4}$ in. to $^{1}/_{2}$ in. (6.4 to 12.7 mm) compressible pad or space is placed directly beneath the shelf angle allows for minor movements. This accommodates shrinkage of concrete and wood frame buildings, as well as long term brick expansion.

The horizontal leg of the angle must extend to support at least two-thirds of the veneer thickness. Angles must also support the veneer without excessive deflection or rotation. Provide a more substantial angle or a built-up structural steel member if deflections exceed about $\ell/600$ to $\ell/720$. In its Technical Note No. 28B *Brick Veneer/Steel Stud Walls* (December 2005), the Brick Industry Association recommends that the deflection of the angle be limited to the lesser of $\ell/600$ or 0.3 in. (7.5 mm) and the rotation of the angle be less than $^1/_{16}$ in. (1.6 mm).

Shelf angles over openings may be supported by bearing on adjoining masonry, subject to building code provisions. In this case, the shelf angle would act as a simply supported beam where the maximum deflection and moment would occur at the midspan.

Shelf angles can also be supported by bolting directly to the structural backup. This is particularly advantageous since the angle size can be minimized by installing bolts at a relatively close spacing, such as the stud spacing. This also reduces the overall deflection of the angle so that only the deflection of the projecting angle leg may be critical. Shelf angles at floor levels must be supported in this manner so that the load from the angle is not transferred to the veneer below.

Avoid attaching heavily loaded shelf angles to steel studs since the substantial bending moment created by the large load and eccentricity could overstress the relatively flexible studs. Any loads imposed upon the studs must be considered in the design of the stud.

When installing shelf angles, a clear space of at least $^1/_2$ inch (12.7 mm) must be left at the ends of each angle. This allows the steel angle to expand without damaging the surrounding masonry.

Similarly, space should be provided between the ends of steel shelf angles at floor lines so each angle can expand and contract independently.

2.7 FLASHING

Flashing materials have also been used increasingly to cover the structural backup, particularly in wood and steel stud construction. Typically these membranes were made from rubber-like plastic membranes or 2 layers of asphalt impregnated building paper.

In anchored masonry veneer, flashing is required directly above all shelf angles, concrete foundations and over door and window openings to interrupt the downward flow of water.

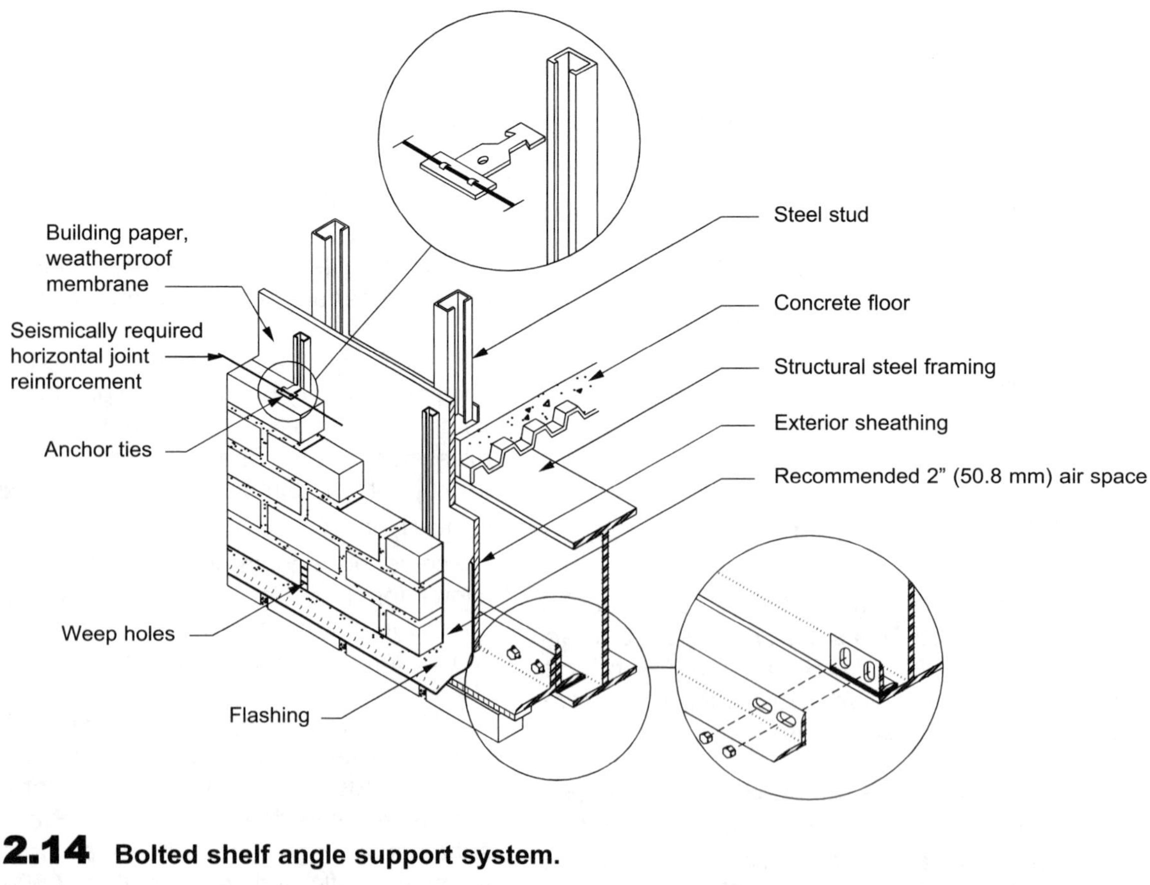

FIGURE 2.14 Bolted shelf angle support system.

Flashing must slope toward the exterior face of the veneer so the water can be shed through weep holes. To prevent the expelled water from dripping down the face of the masonry, the flashing should generally extend beyond the face and terminate in a drip edge. Otherwise, water may re-enter the masonry or cause staining.

Flashings must be continuous along the length of the veneer with sufficient lap joints to prevent moisture from migrating under the flashing. Where flashing terminates at the ends of shelf angles, end dams must be constructed to direct the flow of moisture outward.

Plastic membranes designed specifically for flashing are available. The membranes are thicker than previously manufactured and are reinforced, and resistant to ultra-violet light.

Historically most flashings were made from 10 to 20 ounce per square foot (3.05 to 6.11 kg per square meter) copper or lead sheeting. These materials perform quite well but their use has steadily decreased because of the initial cost and the potential for staining and galvanic corrosion. Advantages of such materials include their availability in a variety of preformed shapes, the ease which they may be worked with and their record of good performance.

Sheet metal flashing of aluminum, galvanized steel and stainless steel are also used, although only stainless steel flashings are generally recommended. Aluminum flashings should be used only with extreme caution since the wet and alkali environment of mortar can corrode the aluminum. Similarly, some galvanized flashings can corrode in fresh mortar and the galvanized coating may crack during bending and handling.

Stainless steel sheet metal flashings provide an excellent corrosion resistant water barrier. Stainless steel is workable yet capable of resisting rough handling at the job-site. It does not stain and is available in several gauges and finishes. When stainless steel sheet metal flashing is used, it should be at least 0.01 in. (0.25 mm) thick and should meet ASTM A167, Type 304.

Asphalt impregnated membrane flashing, also called bituminous fabrics and building felt, is increasingly used as a flashing material since it is more economical than sheet metal flashing and it is easy to install. Unfortunately, asphalt impregnated flashing is less durable than sheet metal flashing and may easily tear during installation.

Accordingly, asphalt systems are seldom used as the sole flashing material and instead are typically used in combination with other flashing materials. In

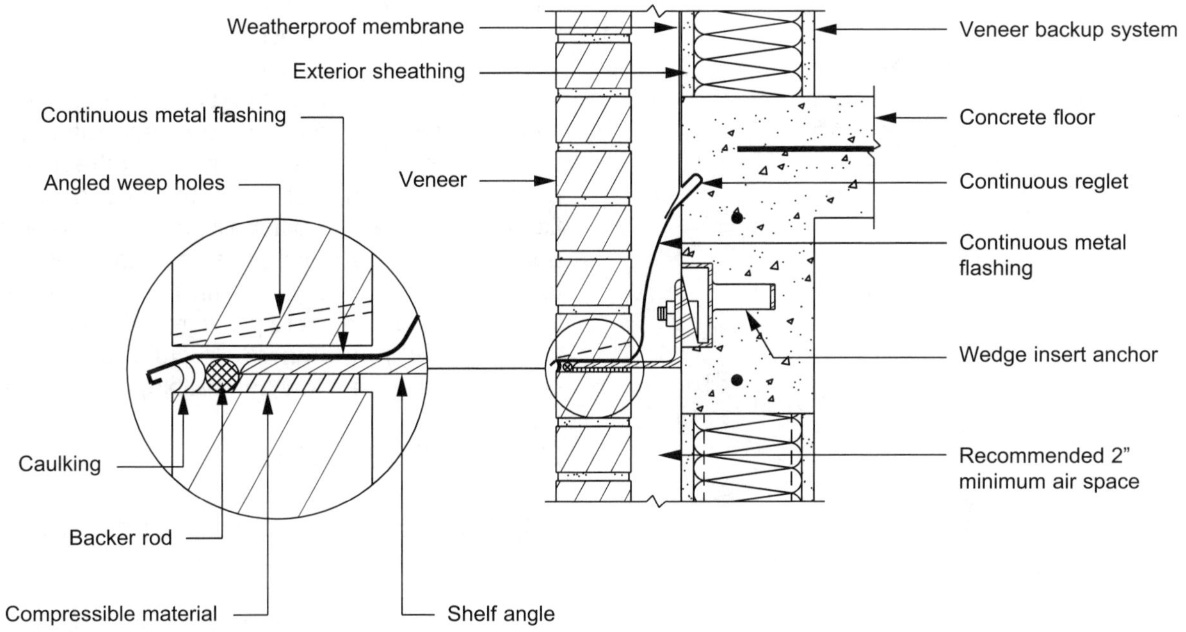

FIGURE 2.15 **Metal flashing system.**

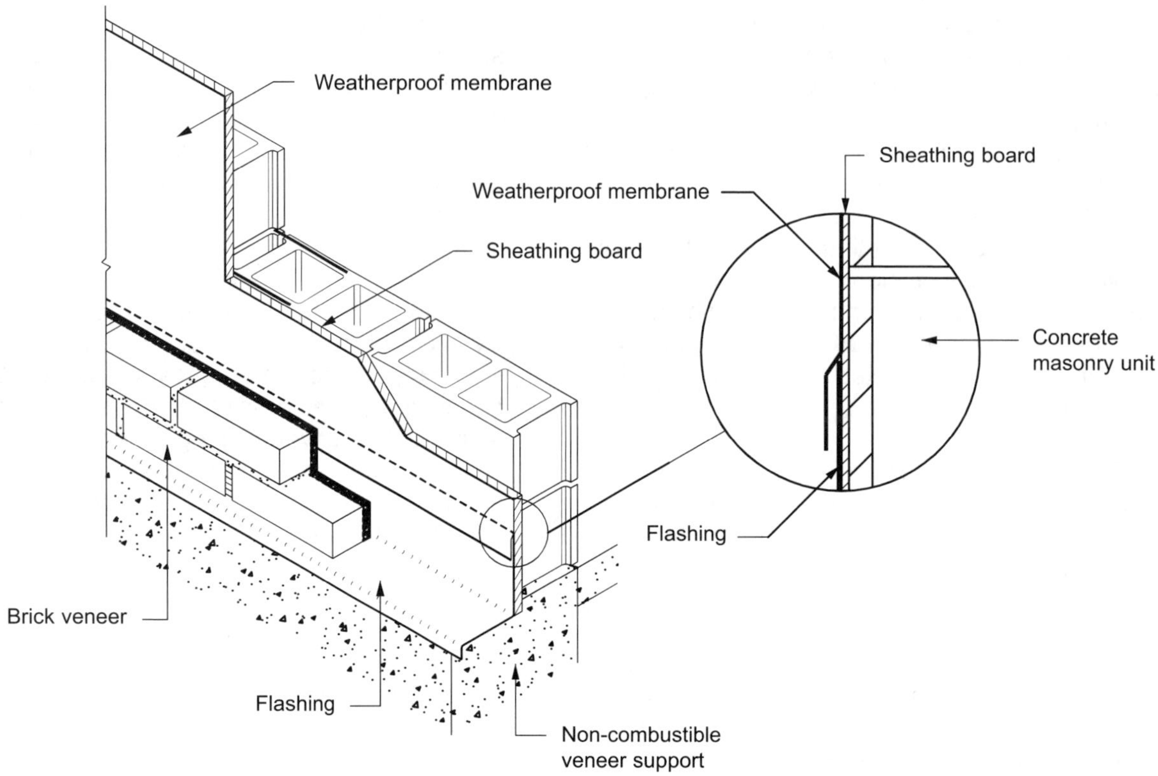

FIGURE 2.16 Masonry flashing material.

such a system, 2 layers of asphalt impregnated sheets such as 30 lb. (13.6 kg) felt are installed over the structural backup so that all seams overlap at least 6 in. (152 mm) and so that the 2 layers are offset by half the sheet width. These layers overlap more resistant flashings at shelf angles and supports.

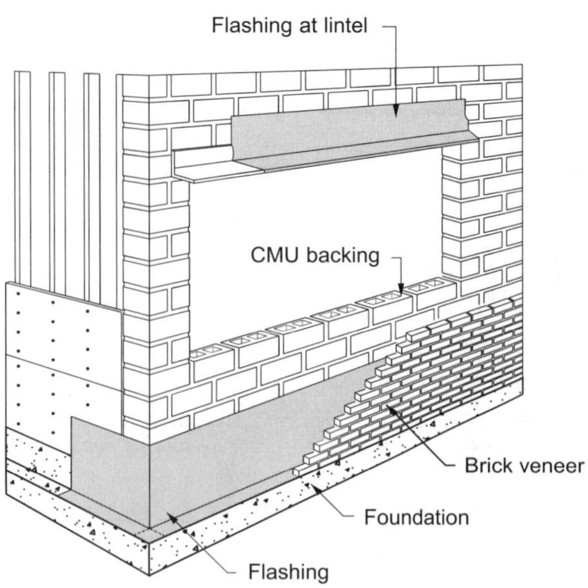

FIGURE 2.17 Adhesive flashing system.

Plastic and rubber-like flashings can perform well if properly designed and installed with care. They are resilient, durable and resistant to corrosion.

There are many plastic flashing materials available. Always consult with reputable manufacturers and obtain test reports on the ultraviolet light resistance and durability of the proposed plastic flashing materials. Additionally, to ensure satisfactory performance, use plastic flashing materials from 20 to 40 mils thick (0.50 to 1.0 mm). Also be aware that low permeability membranes may allow condensation when applied to the entire structural backup. Use a non-permeable, not low permeable, membrane.

2.8 WEEP HOLES

Weep holes are installed above flashing to allow water to readily exit the masonry cavity.

Weep holes are typically placed approximately 24 in. (610 mm) on center, although, the type of weep holes and the particular weather conditions may alter this spacing.

Possibly the easiest form of weep hole is the open head joints. Water can readily drain out of the open head joints which also serve as vents in the wall, allowing air circulation through the cavity. To prevent insects or vermin from entering the open head joints and infesting the cavity, vinyl vents, plastic grids or sheet metal devices can be placed in the open head joint to allow drainage and evaporation.

FIGURE 2.18 **Weep vent.**

One quarter or $^{3}/_{8}$ inch (6.4 or 9.5 mm) diameter plastic tubes can also be used as weep holes. These tubes are installed in the head joints at a slight slope to drain freely. They are less noticeable than open head joints but may have the problem of clogging. Mortar droppings and other debris can lodge in the tube, thus rendering the tubes ineffective. To prevent clogging, some contractors place a few inches of pea gravel or a special mortar screen in the cavity between the veneer and the backup. Also, the plastic tubes may be removed when the section of veneer is complete. The void made by the plastic tube will allow free passage of water to the building exterior.

Cotton, hemp and nylon wick ropes about $^{1}/_{4}$ in. to $^{3}/_{8}$ in. (6.4 to 9.5 mm) in diameter have been used to drain moisture from cavity and veneer walls. Water in the cavity is absorbed by the material and wicked to the exterior where it evaporates. This can be a very slow process therefore the wick ropes should be placed approximately 16 in. (406 mm) on center.

These wicks should be installed so that at least 6 in. (152 mm) of each wick rope penetrates into the cavity. This provides a large surface area for water to be absorbed into the wick despite the possibility of a few mortar droppings.

In time, cotton wick ropes rot, ideally leaving clear drainage holes. Nylon and hemp are also available for use as weep hole material.

Another type of weep hole is formed with $^{3}/_{16}$ to $^{3}/_{8}$ in. (4.8 to 9.5 mm) diameter oiled rope or tube which is installed in the head joints of the veneer. After construction of the veneer is complete, the ropes or tubes are removed to form clear weep openings.

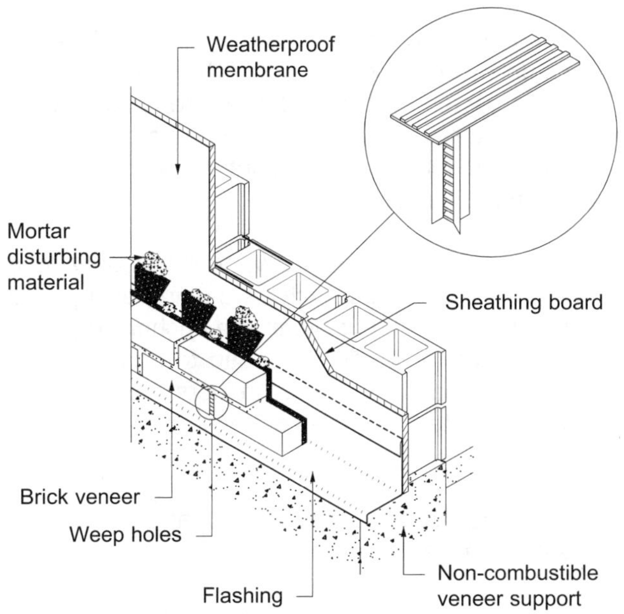

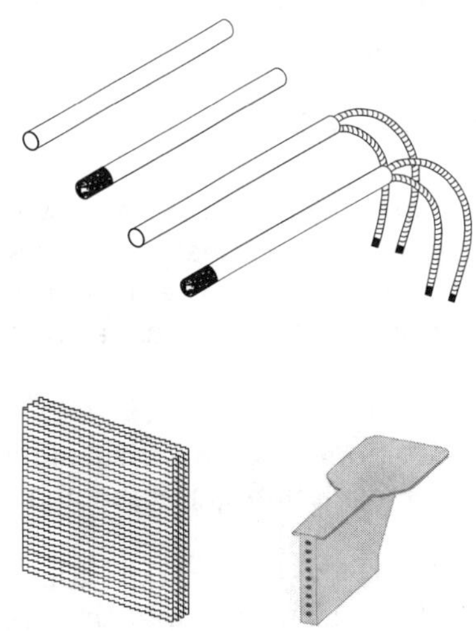

FIGURE 2.19 **Weep hole device.**

2.9 CONNECTORS

When securing the veneer to the backup system, shelf anchor or foundation, numerous types of connectors can be used in masonry construction and the type utilized depends on the particular application, local practice, requirements and availability.

There are three main categories of connectors, and since industry terminology is often ambiguous, definitions will be given.

Wall ties: Used to tie two wythes (widths) of masonry together, as would be the case for cavity walls. Wall ties are commonly manufactured from wire formed into either a square tie or horizontal joint reinforcement as shown in Figure 2.20.

Anchors: Used to secure veneer masonry to a supporting structure such as a stud backup wall or a structural steel column. An anchor may be a flat-type metal fabrication used to attach to the selected backup system. Anchors may also be wire fabricated. Samples of anchors are shown in Figure 2.21.

Fasteners: Used to connect an appliance such as a sign or a fixture to masonry.

Only wall ties and anchors are directly related to anchored veneer construction. Fasteners should be installed in strict accordance with local codes and the manufacturers instructions.

2.9.1 WALL TIES

Wall ties are used to connect two wythes of masonry together and are designed and sized based on prescriptive standards. The ties are sized and spaced to provide adequate capacity in resisting applied loads.

In no case should the tie be substandard to the most stringent minimum requirement that applies.

The 2006 IBC Code Section 1405 provides tie requirements based on the type and application of masonry veneer. Additional veneer tie detail requirements are contained in the 2005 MSJC Code, Chapter 6.

The veneer system must also permit differential movement between the veneer and the backup.

Square ties and horizontal joint reinforcement are the most common wall ties, and are used in conjunction with structural masonry backup.

Additionally, wall ties must be protected to resist corrosion as stated in both the MSJC Code and MSJC Specification.

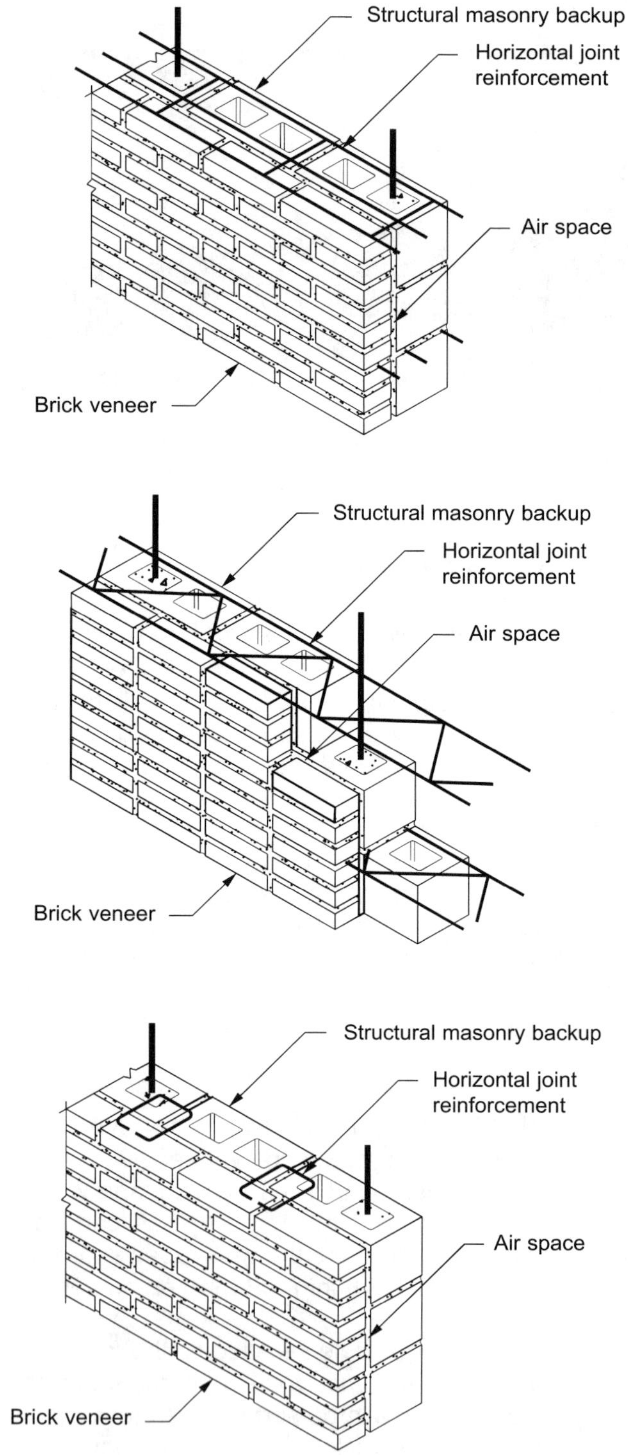

FIGURE 2.20 **Typical wall tie systems.**

Wall ties, sheet metal anchors, steel plates and bars, and inserts exposed to earth or weather, or exposed to a mean relative humidity exceeding 75 percent shall be stainless steel or protected from corrosion by hot–dip galvanized coating or epoxy coating. Wall ties, anchors, and inserts shall be mill galvanized, hot–dip galvanized, or stainless steel for all other cases.

2.9.2 ANCHORS

Different types of anchors are available and when properly designed and fabricated, work efficiently. Selection flexibility should be considered unless there are specific design requirements that dictate the type of anchor required.

Manufacturers are helpful in providing selection assistance and product data to ensure that the anchor will provide a quality connection.

New Zealand, which is an active seismic region, uses a significant amount of masonry veneer. An effective and popular brick anchor in that region is the La Pallé flexible tie connector.

As shown in Figure 2.23, this tie is excellent for transferring the lateral loads to the backup system while allowing for relative movement in the horizontal and vertical directions with the backup system.

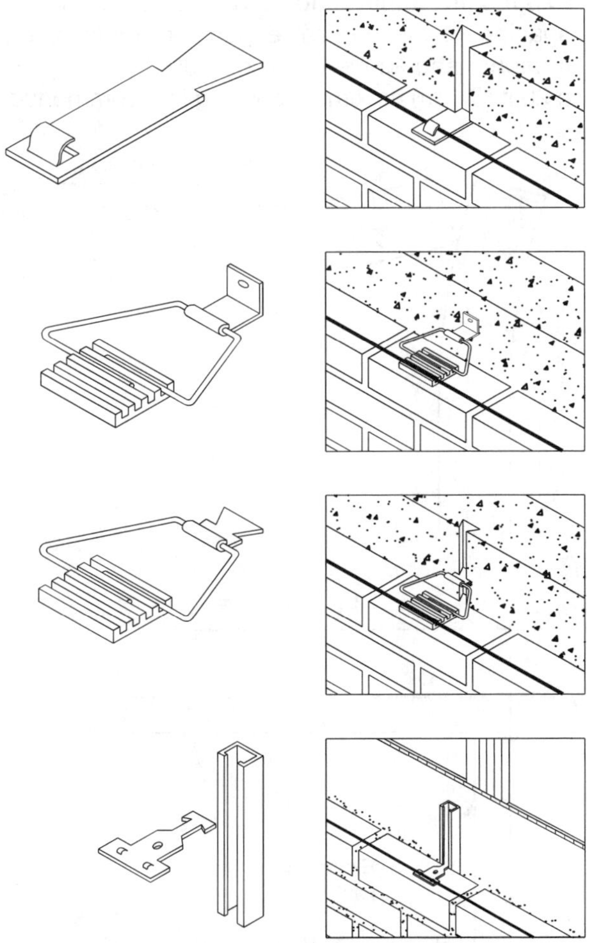

FIGURE 2.21 **Typical seismic veneer anchors.**

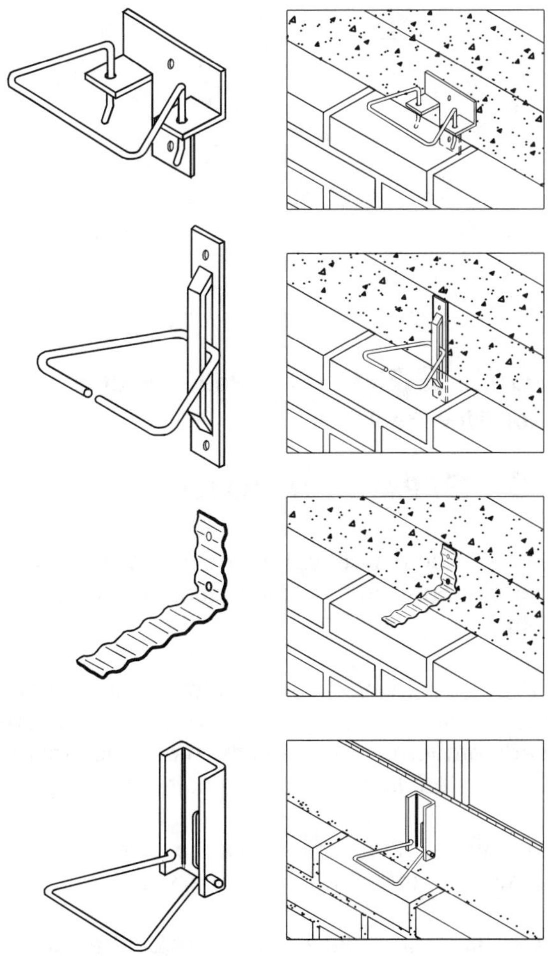

FIGURE 2.22 **Typical non-seismic veneer anchors.**

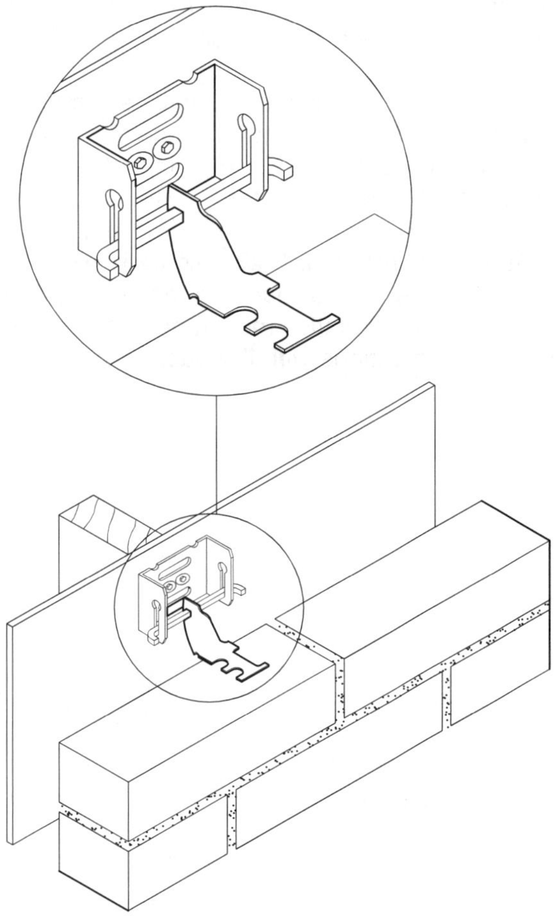

FIGURE 2.23 **La Pallé flexible veneer anchor (New Zealand).**

2.10 EXPANSION JOINTS

Any successful veneer system design will consider expansion and contraction within the entire system.

The need for vertical expansion joints depends on the climate conditions in which the structure is located; temperature change; type of structural frame and the materials involved in construction.

Additional attention should be given to vertical expansion joints in the following areas:

Corners – In order to avoid buckling as depicted in Figure 2.24 some type of vertical expansion joint should be placed in close proximity of the corner.

A vertical expansion joint can be placed at the corner of the veneer and this may be accomplished by following a continuous seam up through a series of attached vertical and horizontal joints. Normally, this method is cost prohibitive.

Alternately, the vertical expansion joint can be placed within ten ft (3.0 m) of the corner. The distance from the corner on the adjacent wall would be the maximum calculated space between joints, less the distance on the adjacent wall.

Discontinuities – When the veneer encounters any discontinuity, a vertical expansion joint should be provided. Examples of changes are changes in height, thickness, and presence of any large openings, dissimilar materials or abutment to other building elements.

Spacing – With all variables taken into consideration, maximum spacing of the vertical expansion joints should be calculated. The spacing of the vertical expansion joints of clay masonry should never exceed 30 ft (9.1 m), and a 20 ft (6.1 m) maximum spacing is recommended.

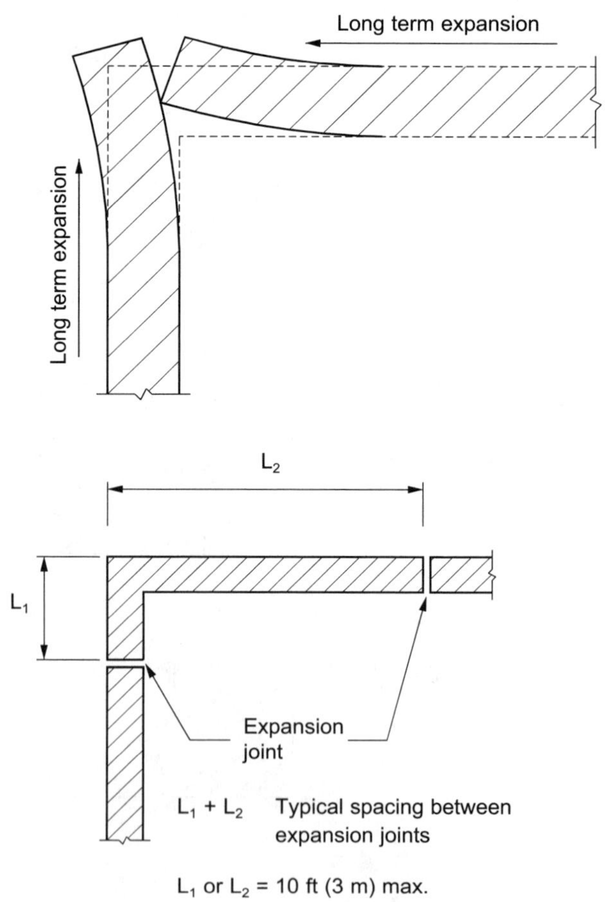

FIGURE 2.24 **Expansion joints at corners.**

3 System Detail Requirements

3.1 General

This section describes masonry veneer systems in detail along with specific requirements contained in the 2006 IBC and the 2005 MSJC Code. Sections and details are provided which show the minimum code requirements along with recommendations based on design principles and standard construction methods.

Minimum thicknesses for various types of wall coverings are contained in IBC Table 1405.2. The given thicknesses are based on weather protection for the material and application. Note that the requirement for adhered masonry veneer is only 0.25 in. (6.4 mm), whereas the anchored masonry veneer must have a minimum thickness of 2.625 in. (66.7 mm). This is based on the concept that adhered veneer is applied to a system that already considers water management, such as a plaster substrate. Anchored veneer, on the other hand is the first line of defense for moisture penetration. The minimum unit thickness of 2.625 in. (66.7 mm) provides adequate surface area for the mason to place the units effectively.

IBC TABLE 1405.2 (Partial)
MINIMUM THICKNESS OF WEATHER COVERINGS

COVERING TYPE	MINIMUM THICKNESS (IN)
Adhered masonry veneer	0.25
Anchored masonry veneer	2.625
Marble slabs	1
Precast stone facing	0.625
Stone (cast artificial)	1.5
Stone (natural)	2
Terra cotta (anchored)	1
Terra cotta (adhered)	0.25

3.2 Veneer Requirements

IBC Chapter 14, *Exterior Walls*, includes masonry veneer and the associated prescriptive requirements. Additionally, IBC Chapter 14 references MSJC Code Chapter 6 for other specific requirements associated with masonry veneer.

The MSJC Code requires that the veneer backing system be able to resist water penetration. More specifically, any sheathing must be covered with a water-resistant membrane, such as 15 or 30 pound felt, unless the veneer backing uses water-resistant sheathing with sealed joints.

Anchored veneer requires flashing and weep holes at the base of the wall, or where any lintels support the veneer above. Maximum spacing of the weep holes is 33 in. (838 mm) which allows for four 8 in. (204 mm) long brick between weep holes, with an extra inch for tolerance. The weep holes are to be a minimum of $^3/_{16}$ in. (4.8 mm) in diameter. If using a plastic tube for a weep hole, the inside diameter of the tube must be $^3/_{16}$ in. (4.8 mm). Alternate devices, such as weep vents, are acceptable if they provide no less passage area than the $^3/_{16}$ in. (4.8 mm) diameter weep hole.

Typically, any veneer attached to a wood stud system or a cold-formed steel framing system is limited to 30 feet (9.1 m) above the noncombustible foundation. This height limitation is increased to 38 feet (11.6 m) for a gable condition.

The restriction on wood framing can be attributed to fire safety. In the event of a fire, the veneer could collapse if the wood stud frame backup disintegrates and a fall of a system in excess of 30 feet (9.14 m) could be catastrophic. There are also considerations of differential movement, that is, a wood frame backup shrinks over a long period of time, whereas, a brick cladding will expand over a long period of time.

Fire protection is considered in the noncombustible systems, and masonry veneer is not a hazard in a moderate fire.

Plain gypsum lath or gypsum board should never be used as an exterior material on the backup system.

3.2.1 DEFINITIONS

IBC Chapter 14 contains definitions that are specifically related to masonry veneer.

IBC Section 1402 (Selected Definitions)

1402.1 General. The following words and terms shall, for the purposes of this chapter and as used elsewhere in this code, have the meaning shown herein.

ADHERED MASONRY VENEER. Veneer secured and supported through the adhesion of an approved bonding material applied to an approved backing.

ANCHORED MASONRY VENEER. Veneer secured with approved mechanical fasteners to an approved backing.

BACKING. The wall or surface to which the veneer is secured.

EXTERIOR WALL. A wall, bearing or nonbearing, that is used as an enclosing wall for a building, other than a fire wall, and that has a slope of 60 degrees (1.05 rad) or greater with the horizontal plane.

VENEER. A facing attached to a wall for the purpose of providing ornamentation, protection or insulation, but not counted as adding strength to the wall.

3.2.2 INSTALLATION

The IBC contains prescriptive guidelines for the installation of masonry veneer. The designer has the option of following the prescriptive guidelines, or to design the masonry veneer in a manner acceptable to the building official.

The 2006 IBC Section 1405.6 (stone veneer) and Section 1405.8 (terra cotta) give prescriptive detail requirements. The code provisions of the IBC and MSJC allow for rational design, which means that the veneer can be designed by engineering calculations, notwithstanding the prescriptive requirements contained in the code. An example of this is masonry veneer designed and constructed as reinforced panelized systems.

Dimension stone, thin cut stone panels, typically installed as cladding on high rise buildings, is not included in the provisions of the 2005 MSJC Code. Dimension stone systems should always be engineered.

As defined, veneers are not a structural part of the building system. The veneer is only to hold its own weight and any lateral loading is to be transferred to the backing.

Surfaces to which veneer is attached shall be designed to support the additional vertical and lateral loads imposed by the veneer. One important design aspect is the deflection of the backup system under imposed lateral loads. Excessive deflection may allow cracking in the masonry veneer.

Figure 3.1 shows a typical load flow from the anchored veneer to the backup system. The veneer system supports only the weight of the veneer, which is transferred to the noncombustible foundation, or to the backup system through the shelf angles.

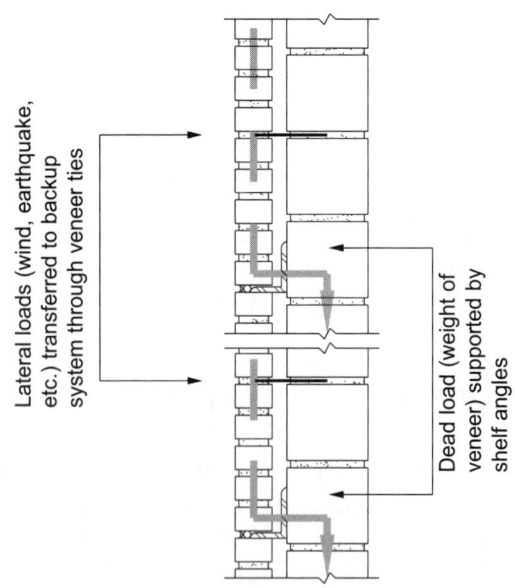

FIGURE 3.1 Load flow of anchored veneer system to backup system.

The code requires that lateral loads, such as wind and seismic loads, be transferred to the backup system through the attachments. In the case of adhered veneer, the adhesive material applied to the back of the veneer transfers the loads. With anchored veneer, the load is transferred through mechanical systems such as wall ties or anchors.

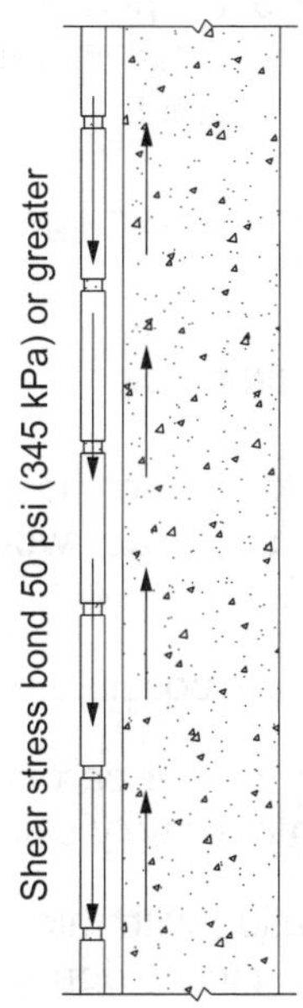

FIGURE 3.2 **Adhered veneer shear.**

In addition to lateral load transfer, consideration must be given for differential movement between the veneer and the backup system. This movement is usually accommodated by the supports. Differential movement is caused by many factors including temperature changes, shrinkage, creep and deflection.

One guideline is that anchored veneer should be designed to resist horizontal forces, such as wind or earthquake, equal to twice the weight of the veneer. This provision is taken from the Uniform Building Code requirements and gives a good sense of quantification. If a 4 in. (102 mm) nominally wide brick veneer system weighs 40 pounds per square foot (1.916 kg/m^2) and has one brick tie for every two square feet, then the tie should be able to withstand 160 pounds (72,600 kg) force in tension or compression. Careful attention must be given to the entire system, especially the connection of the metal tie to the veneer system on one side, and connection to the backup system on the other side.

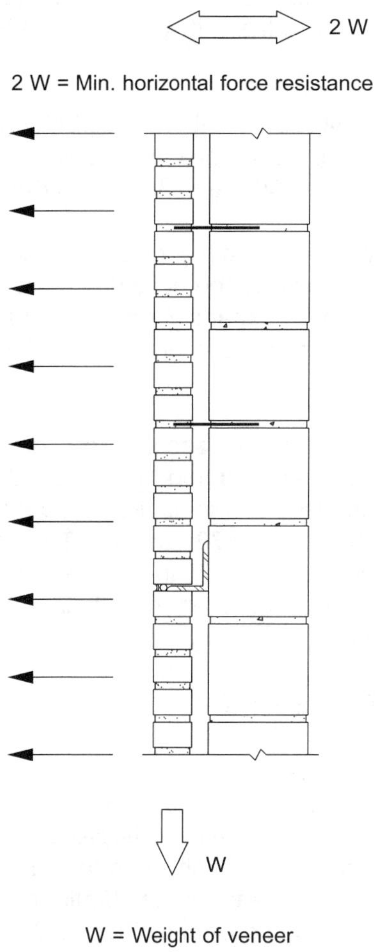

FIGURE 3.3 **Anchored veneer force resistance.**

3.3 ADHERED VENEER

Adhered veneer may be designed and installed without limitation to height and length; however, consideration must be given for brick expansion and building contraction. The code language states that the consideration is to control restrained differential movement between the veneer and backing.

In order to satisfy this requirement, adhered masonry veneer may be applied with vertical and horizontal expansion joints between sections. The vertical distance between horizontal expansion joints would likely be between floors, or approximately 10 ft. (3 m). The horizontal distance between vertical joints should be compatible with the building design and ideally not exceed 15 to 20 ft. (4.6 to 6.1 m).

There is also a unit size limitation imposed by the 2005 MSJC Code, Section 6.3.2.1, which states that the thickness of adhered units are not to exceed $2^5/_8$ in. (66.7 mm). The section further requires that the individual units are not to exceed 5 ft^2. (0.46 m^2) in area and not more than 36 in. (914 mm) long in any face dimension.

Section 1405.9.1 of the 2006 IBC limits interior adhered veneer to a maximum self weight of 20 lbs/ft^2 (98 kg/m^2) and the deflection to ℓ/600.

Consideration must be given to brick masonry veneer expansion and the building contraction of the backing system. If expansion tolerances are not part of the design process, then veneer damage may result.

Adhered veneer is secured by adhesion of a bonding material applied over solid backing. Masonry units used in this application are limited to a weight of 15 lbs/ft^2 (73 kg/m^2). This code also requires that the bond of an adhered veneer to its backing must be designed to withstand a shearing stress of 50 psi (345 kPa).

3.3.1 THIN BRICK

Thin brick veneer is increasingly popular in commercial, residential and do-it-yourself markets. The decision to begin using this product was due to marketing research which recognized remodeling and re-veneering areas as substantial markets for a thin clay veneer wall covering. Thin brick veneers are installed using a variety of procedures. Some of the common methods include:

1. Adhering directly to CMU or Concrete.
2. Adhering to Concrete Block Masonry or Concrete with surface preparation.
3. Adhering to a wood or metal stud system with surface preparation.
4. Placing into forms and cast integrally with concrete.
5. Adhering to prefabricated panels, on the job site, or any other selected location.

Thin brick varies in size, color and texture, depending on the manufacturer. Common face dimensions are approximately $2^1/_2$ in. (64 mm) by $7^5/_8$ in. (194 mm) with a thickness of about $^7/_{16}$ in. (11 mm). The thickness may be as much as $^7/_8$ in. (22 mm), which is preferred when a raked mortar joint is desired.

Longer units of $11^1/_2$ in. (292 mm) are available and careful consideration should be given to units of greater length. Long thin brick veneer units may warp in the firing process, which can cause difficulty for the installer and dissatisfaction for the end user.

Thin brick can provide protection to the material over which it is applied and functions as an architectural wall covering. The texture of thin brick units depends on the method of manufacture and the surface treatment used prior to or after firing. The color also depends on the chemical composition and firing process that is used.

There are some advantages and disadvantages of thin brick veneer:

ADVANTAGES OF THIN BRICK

1. Thin brick veneer is more durable than products such as wood, vinyl siding or aluminum.
2. Installation is possible year-round.
3. Thin brick veneer can be applied by workers who are moderately skilled.
4. Prefabrication with thin brick veneer is easily and economically done.
5. Thin brick provides additional sound and fire resistance. Thin brick walls are lighter in weight than conventional masonry veneer.

DISADVANTAGES OF THIN BRICK

1. The durability of thin brick may not be equivalent to conventional brick veneer.
2. The structural properties of conventional brick veneer are not provided by thin brick veneer.
3. Conventional brick veneer has higher sound and fire resistance ratings than thin brick veneer.
4. Conventional brick veneer provides greater thermal mass than thin brick veneer.

The application of thin brick allows the designer and installer flexibility in detailing finish interfaces with other material. As shown in figures 3.4 and 3.5, special thin brick corner units may be used at corners or the corners may be mitered providing a smiliar effect. Since thin brick are adhered, there is no support requirement at the base, as depicted by Figure 3.6.

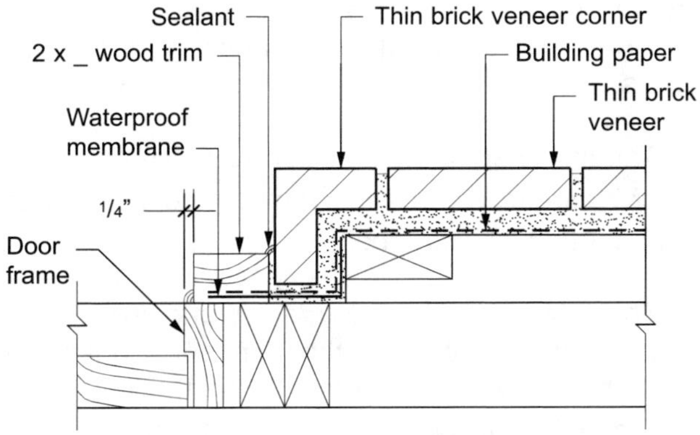

FIGURE 3.4 **Door jamb corner return.**

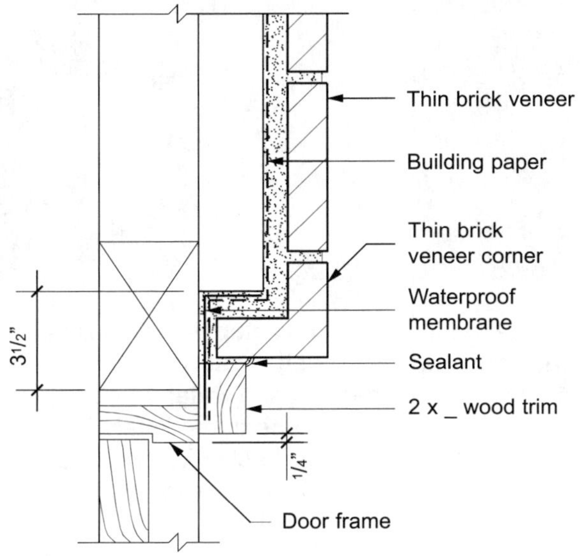

FIGURE 3.5 **Door header corner return.**

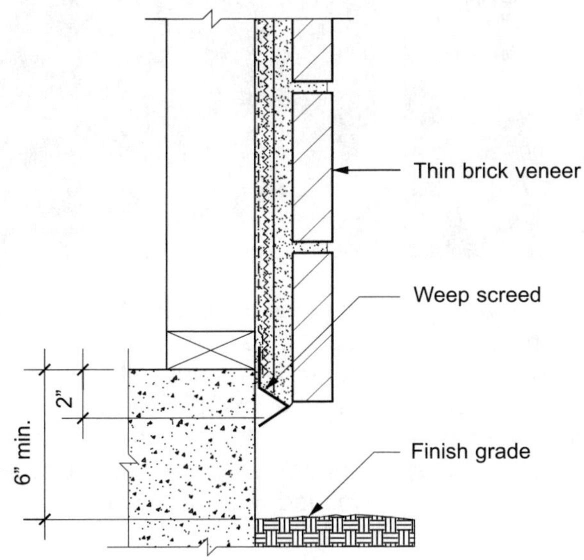

FIGURE 3.6 **Base detail.**

3.3.2 HONEYCOMB STONE

There is a product known as "Light Stone" made by cutting stone to a thickness of about 5 mm, and attaching to aluminum honeycomb material. It can be made in panels as large as 5 ft (1.5 m) by 9 ft 4 in. (2.8 m). The total panel thickness is about 20 mm and the panels are widely used in elevators, since the weight is less than 3 pounds per square foot (15 kg/m^2), thus having minimal affect on the capacity rating of the elevator.

FIGURE 3.7 **Honeycomb stone.**

3.3.3 TERRA COTTA

Terra cotta comes in an infinite variety of shapes and sizes, and most units are anchored.

Adhered terra cotta are thin masonry units, up to $1^1/_4$ in. (31.8 mm) in thickness, including back ribs. The back ribs are necessary in order to provide a secure bond with the mortar.

Adhered terra cotta should never be applied to structural masonry since the masonry units will significantly expand and contract during normal moisture retention and drying cycles. A waterproof membrane should first be applied to the concrete masonry surface, then an expanded metal lath wire mesh and a scratch and brown coat of plaster applied prior to the adhered installation of terra cotta.

3.3.4 ADHERED VENEER INSTALLATION

Adhered masonry veneer is typically installed by one of two common methods. The traditional method is called the thick set method. More recently, the thin set method has become quite popular and uses advanced technology to address some of the traditional shortcomings of adhered veneer installation.

The thick set method may be used on backing of masonry, concrete, wood or steel stud framing. When applied directly to concrete or masonry, the wire lath may be eliminated if the surface is heavily scarified or roughened by sandblasting or other means.

The thick set method may use a setting bed of $^{1}/_{2}$ in. (12.7 mm) to $1^{1}/_{4}$ in. (31.8 mm) where the units are pushed into the setting bed.

The thin set method is more contemporary and uses a latex additive in the thin set mortar. The setting bed is typically $^{1}/_{16}$ in. (1.6 mm) thick and set on a scratch/brown coat of plaster backing.

The procedure is as follows: At the beginning of daily operations, all walls to be faced are drenched with clean water applied through a hose nozzle. And not sooner than one hour before the setting of the veneer, the walls are drenched again.

The veneer to be set is soaked in a vat or tub of clean water for one hour or more just prior to installation.

The next step is the application of a brush coat of neat Portland cement and water only on the wall area to be immediately faced and to the backs of the pieces of veneer to be used.

This brush coat should be watery, not thick and pasty. Immediately over this brush coat spread a mortar coat on both the wall and the thin veneer backs.

It is recommended that no dolomite, magnesium hydrate or processed lime be added to this mortar. The piece of veneer is tapped in place on the wall so that all voids are filled and air pockets eliminated.

The face joints should be of width specified and can be pointed with the same mortar. However, they may be raked out at once and pointed later with a colored mortar. After setting, surface is washed with clean water.

FIGURE 3.8 Adhered veneer installation.

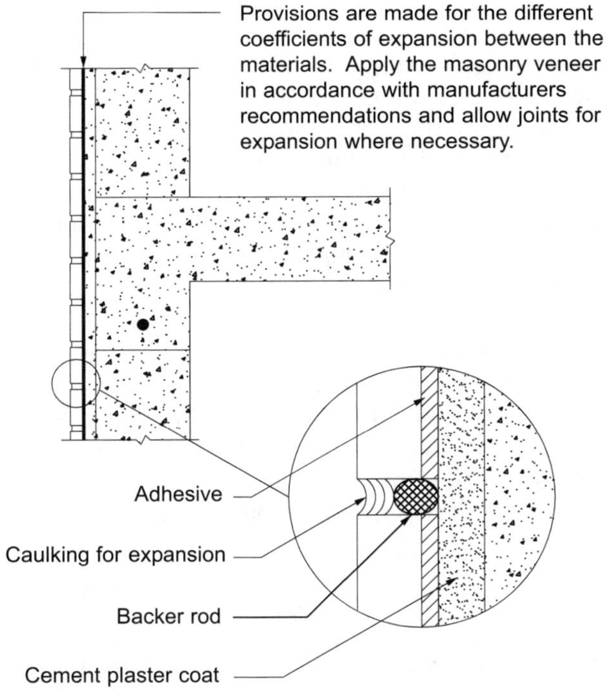

Proper joint allows the concrete backing structure to move, while the caulking compresses. No stresses are transferred to the masonry system.

FIGURE 3.9 Adhered veneer with provision for expansion.

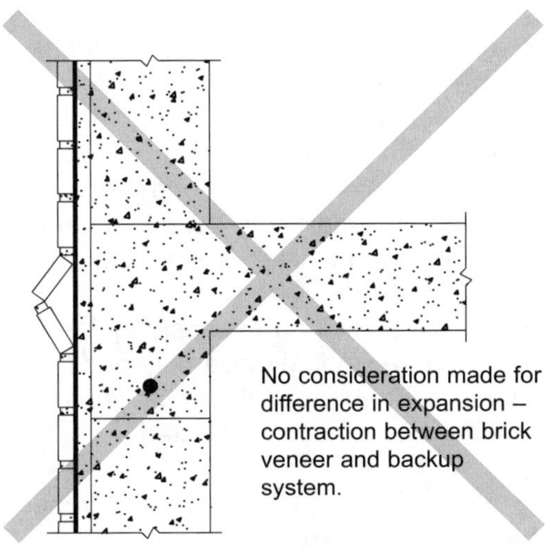

WRONG

FIGURE 3.10 Veneer buckling due to expansion of brick and creep/plastic flow of frame.

Adhered terra cotta is installed by brushing a coat of neat Portland cement and water onto the terra cotta unit and a limited area of wall. The terra cotta unit should be soaked prior to setting.

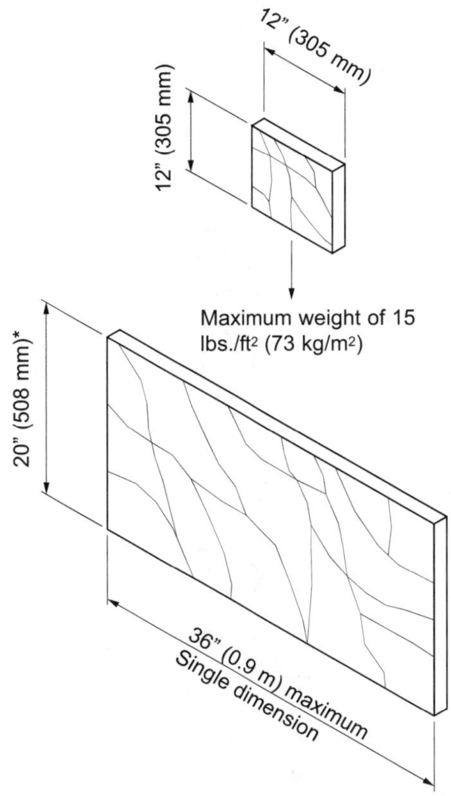

*Maximum dimension in this configuration. Maximum dimensions and maximum area (720 square inches (0.46 m^2)

Adhered veneer units shall not exceed $2^5/_8$ in. (66.7 mm) in specified thickness, 36 in. (914 mm) in any face dimension, nor more than 5 ft^2 (0.46 m^2) in total face area, and shall not weight more than 15 lb/ft^2 (73 kg/m^2).

FIGURE 3.11 Limit of adhered stone dimensions and weight.

Then, spread half the mortar setting bed on the wall and half on the terra cotta unit. Screed both mortar beds with a notched trowel. The total mortar setting bed should be about $^3/_4$ in. (19.1 mm) thick.

Firmly place the terra cotta unit on the mortar setting bed and tap into place with a rubber mallet.

This should assure a full mortar setting bed. Finish joints as appropriate.

The physical characteristics of adhered veneer are such that when set in the recommended manner, it becomes a monolithic unit with the structural wall. Because of these characteristics, both veneer and wall must be noticeably damp at time of setting.

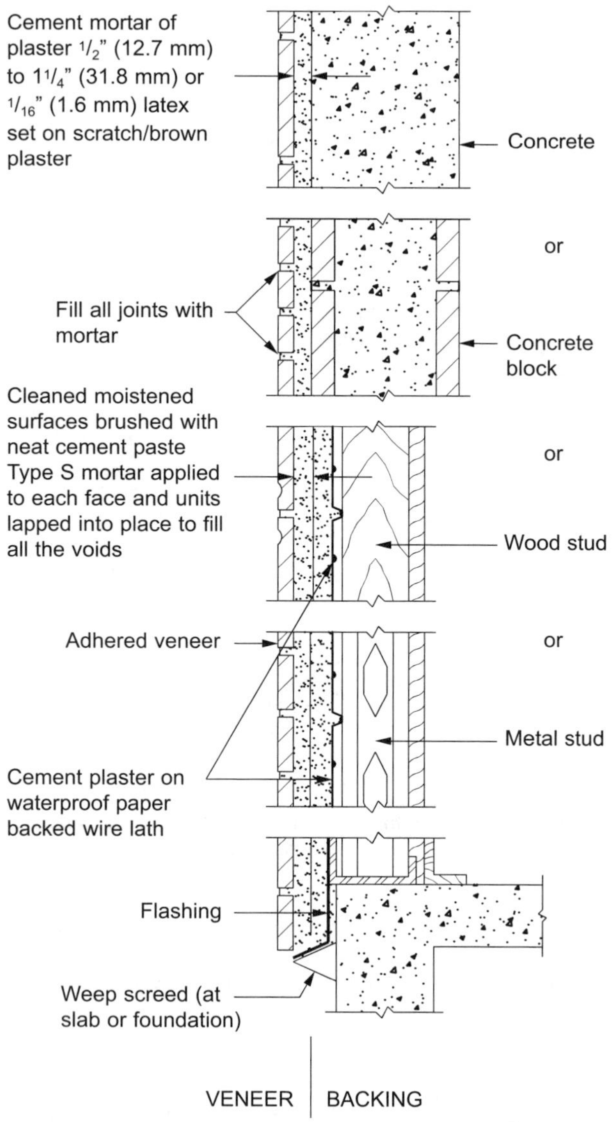

FIGURE 3.12 Adhered veneer wall sections.

3.4 ANCHORED VENEER

2006 IBC Section 1405.5 references 2005 MSJC Code Sections 6.1 and 6.2 for masonry veneer design and installation with some specific prescriptive requirements for stone veneer, slab-type veneer, and terra cotta.

Backing is defined as the wall or surface to which the veneer is secured. The IBC has specific veneer requirements including the materials permitted as the backing to masonry veneer.

Specific height and weight limitations apply for anchored masonry veneer under certain conditions.

For exterior masonry veneer supported by preservative wood treated foundations, the height of the veneer is limited to 18 ft (5.5 m). Also, exterior masonry veneer not exceeding 40 lb/ft^2 (195 kg/m^2) may be supported on wood construction and is limited to a height of 12 ft (3.7 m).

There is also a weight limitation for interior masonry veneer of 40 lb/ft^2 (195 kg/m^2) when used as an interior finish on wood framing.

2005 MSJC Code Table 6.2.2.3.1 lists the maximum height for anchored veneer with backing of wood framing or cold formed steel framing as 30 ft (9.14 m) at the plate line and 38 ft (11.58 m) at the top of a gable. Movement joints allowing for vertical movement directly under shelf angles should be installed.

MSJC Table 6.2.2.3.1 – Height Limit from Foundation	
Height at plate, ft (m)	**Height at gable, ft (m)**
30 (9.14)	38 (11.58)

Since there must be a provision for expansion and contraction at horizontal breaks, or more specifically, directly under the ledger angles, airspace or a compressible expansion joint material must be provided.

For openings, such as doors and windows, where the veneer is not self spanning, noncombustible, noncorrosive lintels shall be used and the deflection of the lintel shall not exceed $\ell/600$ or 0.3 in. (7.6 mm) of the span under the full load of the veneer. For example, the deflection for a 10 ft (3 m) opening shall not exceed 0.2 in. (5 mm).

Figure 3.15 shows an expansion joint detail that will accommodate vertical movement and Figure 3.16 depicts what could happen if vertical movement joint is improperly detailed or constructed.

The area and length of anchored veneer walls shall be unlimited, except as required to control expansion and contraction. This is implied in the 2005 MSJC Code Section 6.1.5, General Design Requirements, which states in Section 6.1.5.3 "Design and detail the veneer to accommodate differential movement".

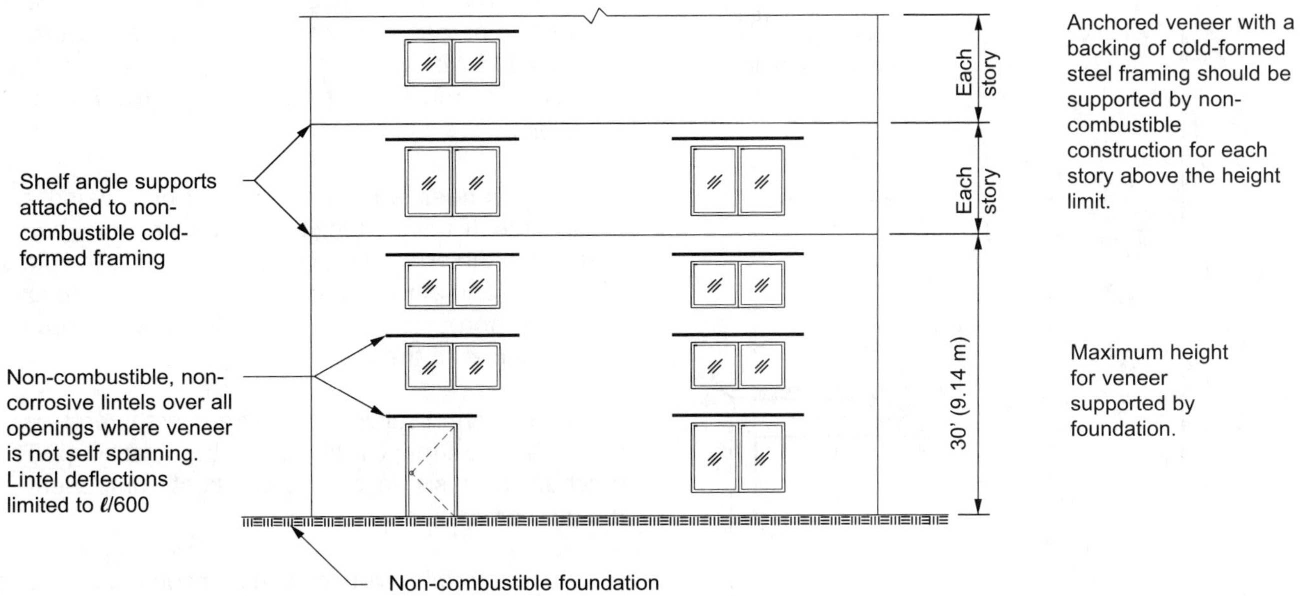

FIGURE 3.13 **Veneer system over 30 feet (9.14 m) in height with a backing of cold-formed steel framing.**

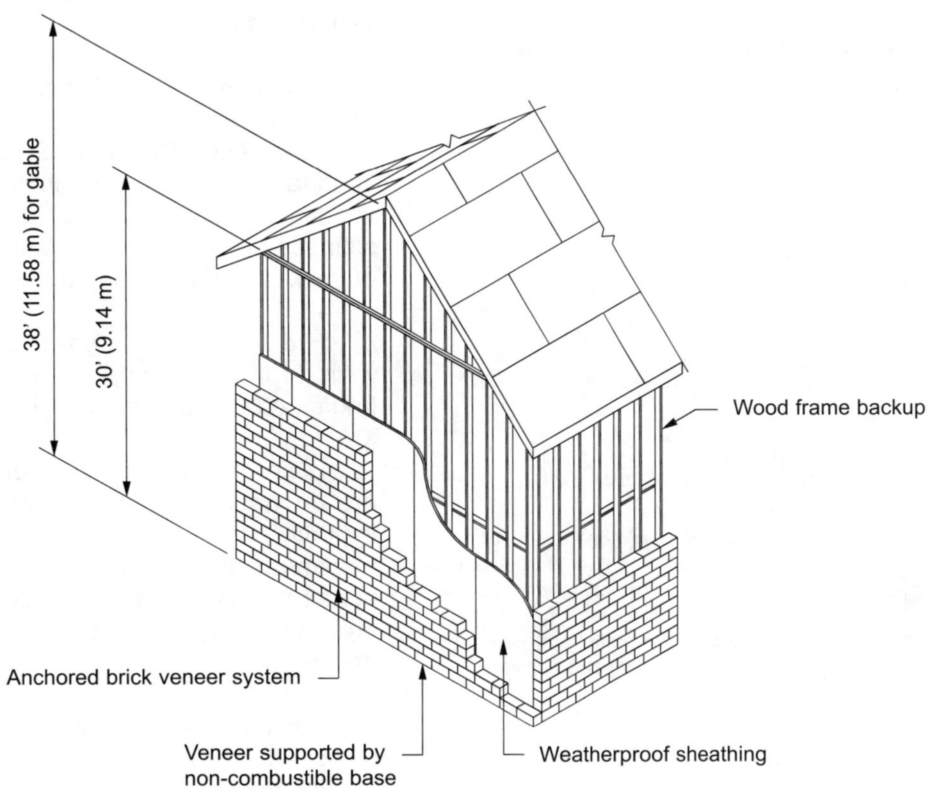

FIGURE 3.14 **Height limit of wood frame backup system.**

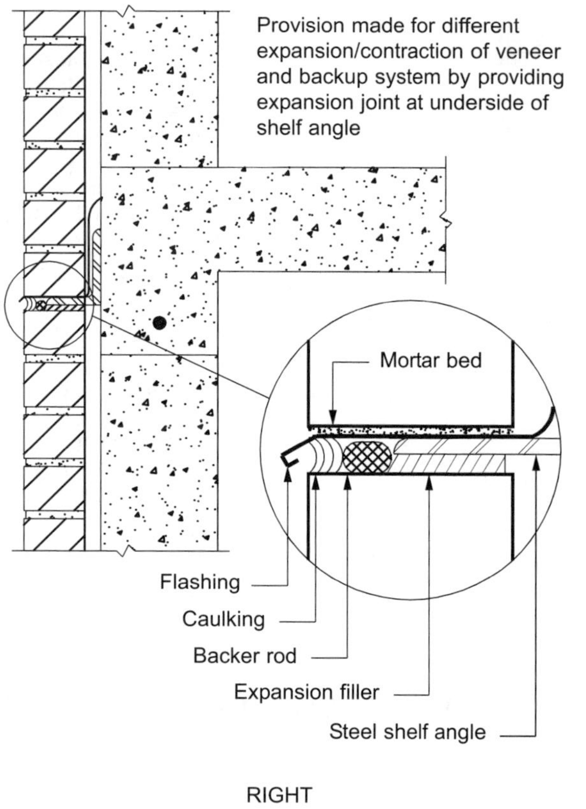

FIGURE 3.15 Anchored veneer with provision for expansion.

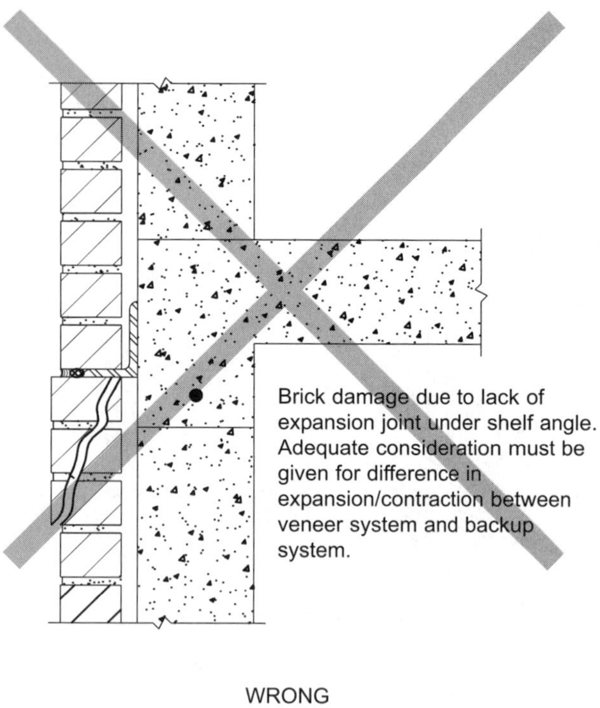

FIGURE 3.16 Anchor veneer with no provision for expansion.

Brick veneer units may be made of clay or concrete. Clay brick has been the most popular masonry material for veneer applications. According to the 2005 MSJC Code Section 6.2.2.4 anchored veneer units must be at least $2^5/_8$ in. (66.7 mm) in thickness.

Brick veneer is commonly used over wood and metal stud framing. Flexible metal anchors permit horizontal and vertical movement parallel to the plane of the wall and at the same time resistance to tension and compression forces in the direction perpendicular to the wall.

The mechanical tie system transfers lateral loads from the veneer cladding to the backing. The mechanical ties can be the weakest component of the veneer system.

Anchored brick veneer construction consists of a nominal 3 in. (76.2 mm) or 4 in. (102 mm) thick exterior brick wythe tied to a backup system with metal ties in such a way that a 1 in. (25.4 mm) minimum clear space is provided between the veneer and the backup system. The brick veneer is supported on the foundation and should not carry any vertical loads, other than the weight of the veneer.

Some properties of brick veneer are:

Durability – Consideration must be given to five important elements that affect the durability of brick veneer, namely, the height of the brick veneer, the stiffness of the backing, the tie system, and the type of brick and type of mortar used.

Fire resistance – Typical anchored brick veneer walls (4 in. (102 mm)) have a fire rating of up to 2 hours (2006 IBC Table 720.1(2), Row 1).

Water resistance – Anchored brick veneer walls are drainage type walls. Walls of this type, which include cavity walls, are recommended where maximum resistance to rain and other water penetration is desired. Careful consideration must be given to the design and construction to permit water drainage.

Insulation – Brick veneer walls can easily be designed to obtain an energy efficient building environment in any type of weather condition.

Acoustics – Brick veneer walls reduce sound transmission. The mass reduces sound transmission by absorbing the energy of the sound vibration.

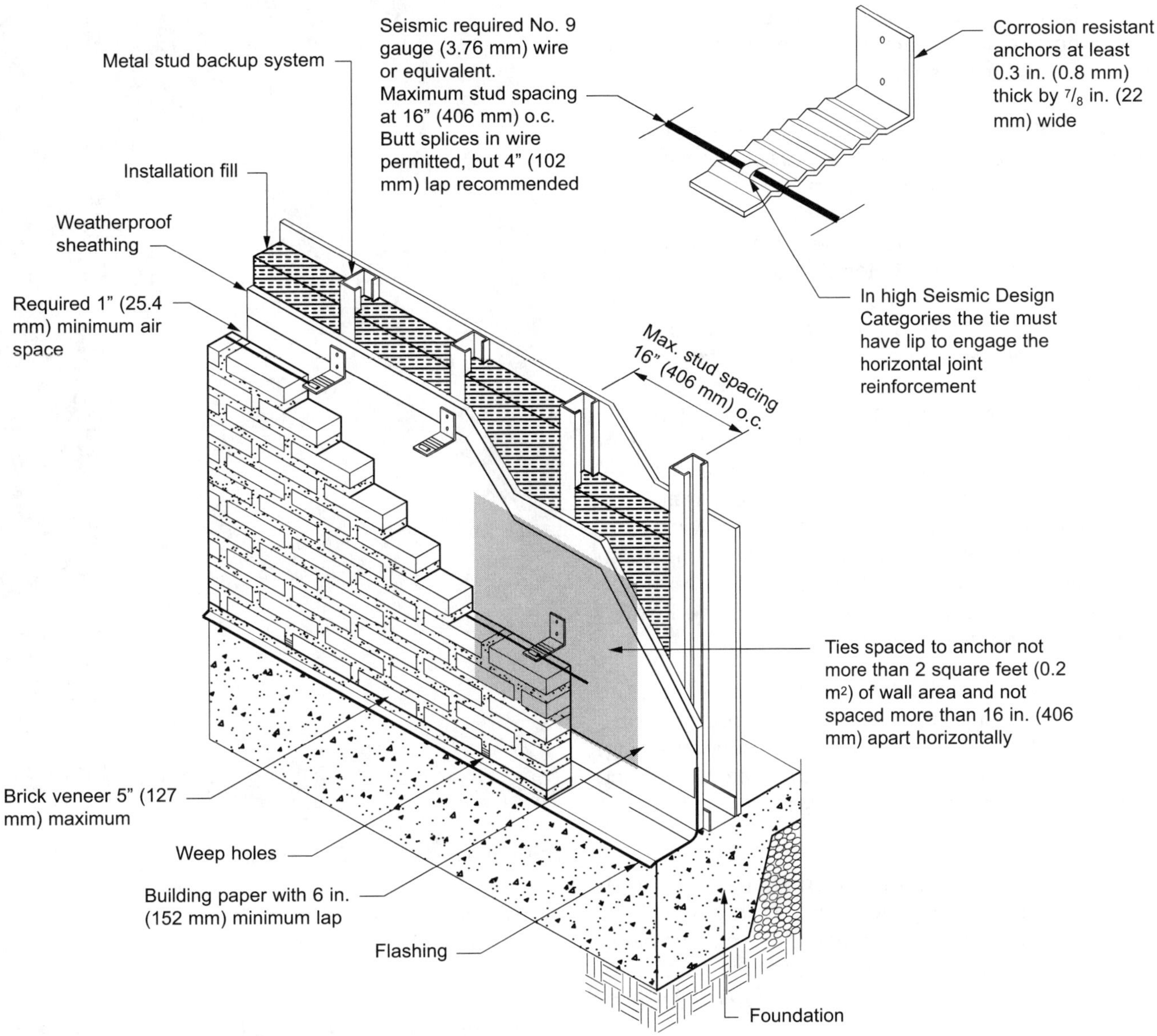

FIGURE 3.17 Seismic anchored veneer tie system.

3.4.1 STONE VENEER

The application of stone as a cladding in construction is a continually evolving process. Stone was first used for shelter because it formed the walls of caves in which early man took refuge from the ravages of the elements.

Some features of stone construction include natural beauty, the subliminal feeling of security and strength that it creates. There are two basic types of natural stone veneer:

1. Rubble or cut stone laid in mortar, and
2. Thin 10 mm to 30 mm stone slabs mechanically or adhesively attached.

Most low-rise residential and small commercial buildings use mortar for the installation of stone.

Successful applications of exterior veneer on various types of structures have made building stone a very recommendable and desirable solution to cladding challenges. The flexibility of the application of building stone has optimized the cost effectiveness of installation. This is especially true with the wide variety of architectural features, such as color, texture and pattern that can be achieved.

The choice of building stone for exterior application is made on the basis of several fundamental criteria: durability, color, surface finish, initial cost and maintenance expense.

Durability is a characteristic of the material to withstand climatological effects, resist atmospheric agents and attack from pollutants.

A distinction should be made between weathering and decay of stone. Not all changes by weathering agents are necessarily undesirable or harmful in that they do not always destroy or materially alter the natural integrity of the stone within a given length of time, although when exposed to the processes of weathering, all material must ultimately decompose or disintegrate. For example, some changes in color may not materially affect the integrity of the stone but may be desirable for architectural or aesthetic effects, giving an appearance of age. Hence, the element of time must enter into the evaluation of weathering in terms of modes and environmental conditions of use insofar as these can be evaluated satisfactorily.

Color is a particularly important feature in architectural design. Contacting a distributor or local marble or stone contractor will assist in determining availability of stone colors. Stone is a natural product and color and characteristics vary from one piece to the next.

Stone surfaces can be finished in a number of ways with smoother and more highly polished finishes yielding more apparent color and veining. Rougher finishes, on the other hand, will cause the color and veining to be less obvious. Typical finishes are:

Abrasive Finish – A flat, non-reflective slightly roughened surface; excellent for exterior use.

Bush Hammered – A corrugated finish, smoother on small surfaces, with interrupted parallel markings depending upon actual finishing: 8-cut (marking not over $^3/_{32}$ in. (2.4 mm) apart), 6-cut (coarser with markings not over $^1/_8$ in. (3.2 mm) apart), and 4-cut (coarsest, with markings not more than $^7/_{32}$ in. (5.5 mm) apart).

Cleft Finish – A random finish accomplished with a splitter or guillotine.

Flame or Thermal Finish – A plane, non-reflective surface with highlighted accents, usually recommended for external and paving use. The surface coarseness varies depending on the grain structure of the granite. This finish is achieved by applying heat which pops the surface.

FIGURE 3.18 **Cleft finish.**

Honed Finish – A smooth finish with little or no shine.

Polished Finish – A reflective finish where color and veining is maximized.

Sandblasted Finish – A non-glossy dull surface appropriate for exterior applications.

Waterblast – A process that dulls the surface but retains the color of material.

The glossier and more highly reflective surfaces require more fabrication and finishing than the less finished surfaces, consequently the closer the stone gets to a polished surface the more expensive the material. An abrasive finish is often the most economical for exterior use.

Other finishes, such as machine tooled, split, cleft or rock-faced, chat or shot sawn or tooled, are also available.

FIGURE 3.19 Stone veneer.

Stone veneer may be applied by using the following methods as described in the 2006 IBC, Section 1405.6.

IBC Section 1405.6

1405.6 Stone veneer. Stone veneer units not exceeding 10 inches (254 mm) in thickness shall be anchored directly to masonry, concrete or to stud construction by one of the following methods:

1. With concrete or masonry backing, anchor ties shall be not less than 0.1055-inch (2.68 mm) corrosion-resistant wire, or approved equal, formed beyond the base of the backing. The legs of the loops shall be not less than 6 inches (152 mm) in length bent at right angles and laid in the mortar joint, and spaced so that the eyes or loops are 12 inches (305 mm) maximum on center (o.c.) in both directions. There shall be provided not less than a 0.1055-inch (2.68 mm) corrosion-resistant wire tie, or approved equal, threaded through the exposed loops for every 2 square feet (0.2 m^2) of stone veneer. This tie shall be a loop having legs not less than 15 inches (381 mm) in length bent so that it will lie in the stone veneer mortar joint. The last 2 inches (51 mm) of each wire leg shall have a right-angle bend. One-inch (25 mm) minimum thickness of cement grout shall be placed between the backing and the stone veneer.

2. With stud backing, a 2-inch by 2-inch (51 by 51 mm) 0.0625-inch (1.59 mm) corrosion-resistant wire mesh with two layers of water-resistive barrier in accordance with Section 1404.2 shall be applied directly to wood studs spaced a maximum of 16 inches (406 mm) o.c. On studs, the mesh shall be attached with 2-inch-long (51 mm) corrosion-resistant steel wire furring nails at 4 inches (102 mm) o.c. providing a minimum 1.125-inch (29 mm) penetration into each stud and with 8d common nails at 8 inches (203 mm) o.c. into top and bottom plates or with equivalent wire ties. There shall be not less than a 0.1055-inch (2.68 mm) corrosion-resistant wire, or approved equal, looped through the mesh for every 2 square feet (0.2 m^2) of stone veneer. This tie shall be a loop having legs not less than 15 inches (381 mm) in length, so bent that it will lie in the stone veneer mortar joint. The last 2 inches (51 mm) of each wire leg shall have a right-angle bend. One-inch (25 mm) minimum thickness of cement grout shall be placed between the backing and the stone veneer.

The two methods contained in the 2006 IBC Section 1405.6 are known as "wired tied" anchored veneer systems. Figures 3.21 and 3.22 demonstrate the systems.

Figure 3.24 shows the periphery dowels which may be required by the Code. These anchors are to be connected to the backing by wire or ties.

As with any anchored veneer system, ties must be capable of resisting tension and compression forces. Since there is no mortar spot bedding requirement, the tie must resist the force, or the system can be designed to place spot bedded mortar to accommodate compressive forces.

FIGURE 3.20 Typical anchored stone veneer application.

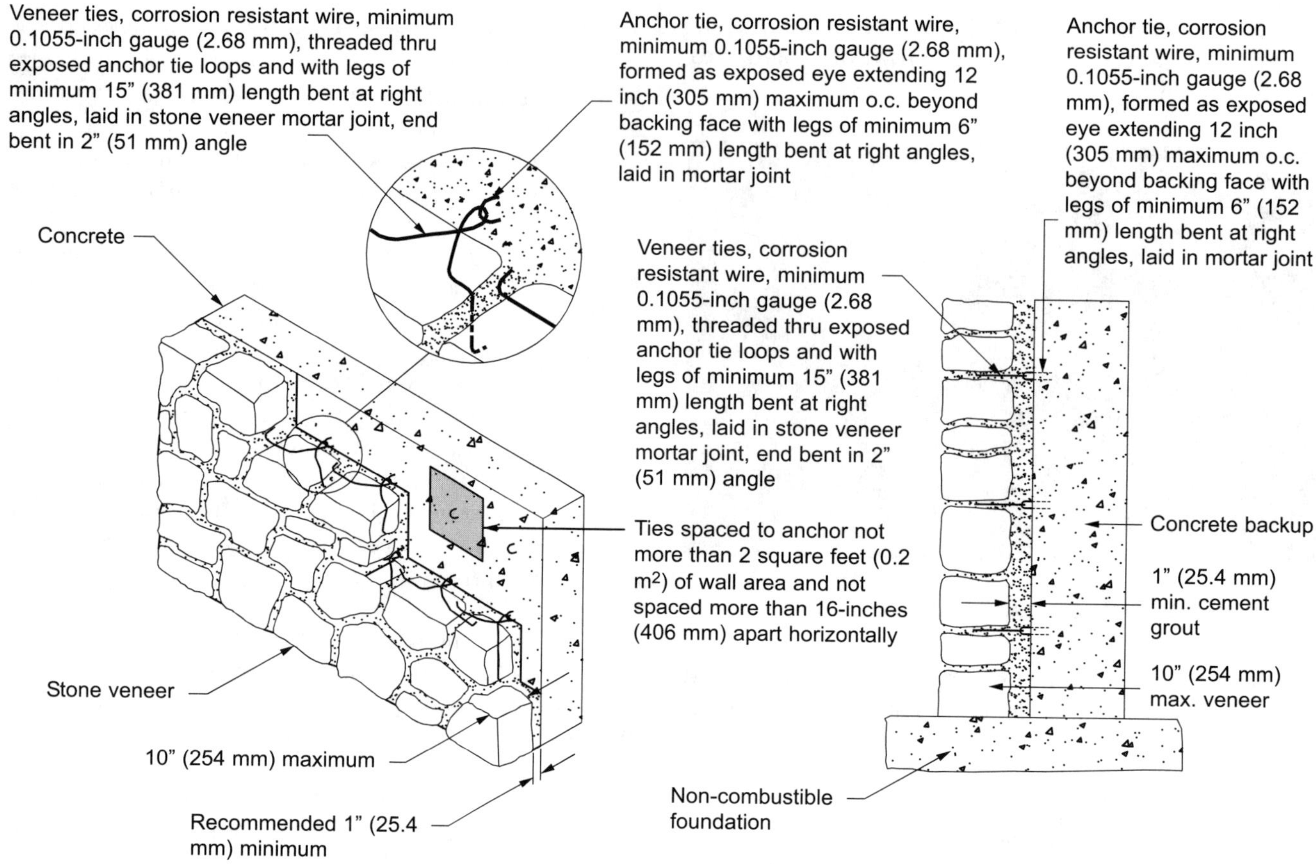

FIGURE 3.21 Anchored system with concrete backup.

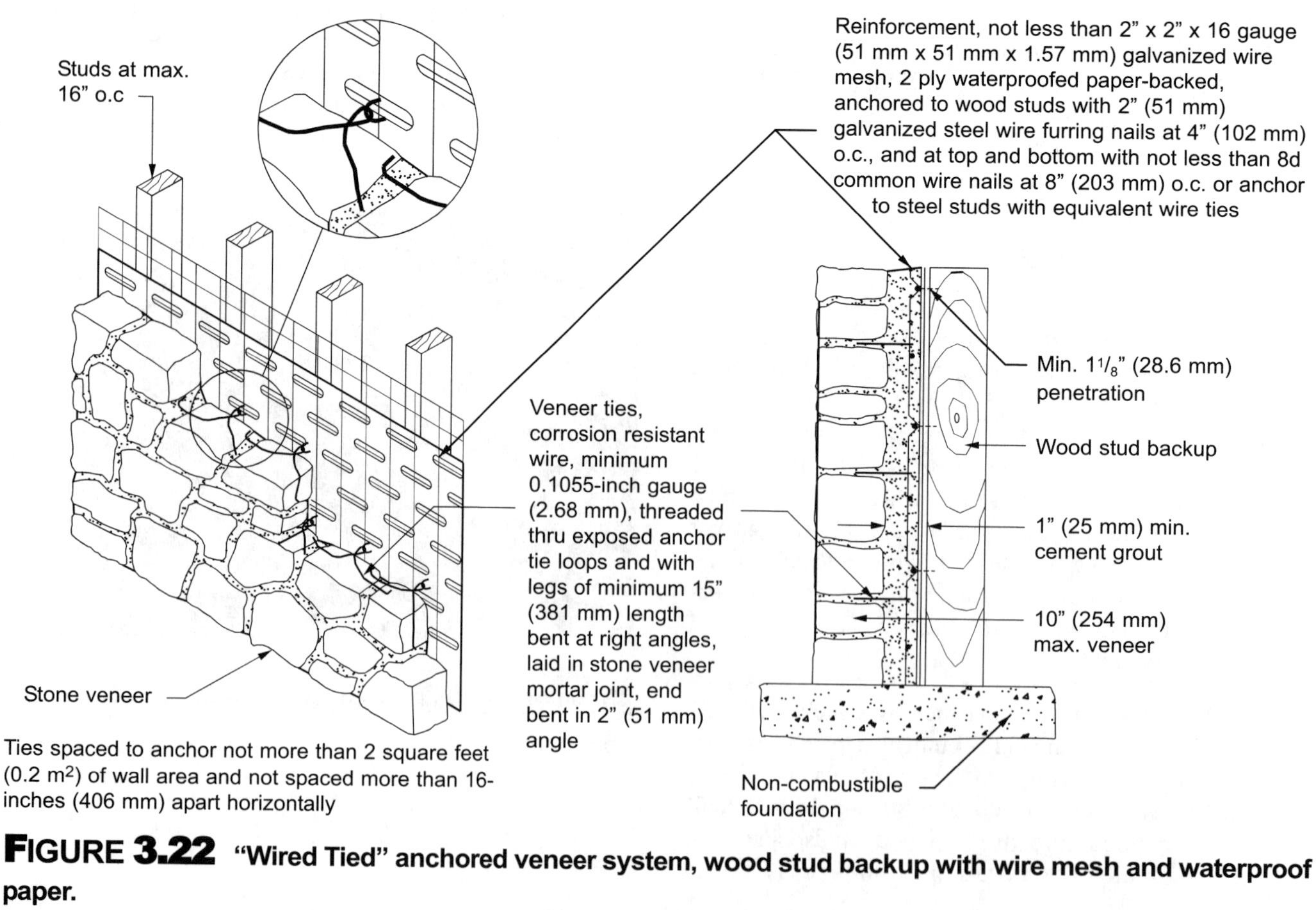

FIGURE 3.22 "Wired Tied" anchored veneer system, wood stud backup with wire mesh and waterproof paper.

IBC Section 1405.7

1405.7 Slab-type veneer. Slab-type veneer units not exceeding 2 inches (51 mm) in thickness shall be anchored directly to masonry, concrete or stud construction. For veneer units of marble, travertine, granite or other stone units of slab form ties of corrosion-resistant dowels in drilled holes shall be located in the middle third of the edge of the units spaced a maximum of 24 inches (610 mm) apart around the periphery of each unit with not less than four ties per veneer unit. Units shall not exceed 20 square feet (1.9 m^2) in area. If the dowels are not tight fitting, the holes shall be drilled not more than 0.063 inch (1.6 mm) larger in diameter than the dowel, with the hole countersunk to a diameter and depth equal to twice the diameter of the dowel in order to provide a tight-fitting key of cement mortar at the dowel locations when the mortar in the joint has set. Veneer ties shall be corrosion-resistant metal capable of resisting, in tension or compression, a force equal to two times the weight of the attached veneer. If made of sheet metal, veneer ties shall be not smaller in area than 0.0336 by 1 inch (0.853 by 25 mm) or, if made of wire, not smaller in diameter than 0.1483-inch (3.76 mm) wire.

FIGURE 3.23 Typical anchored slab-type veneer.

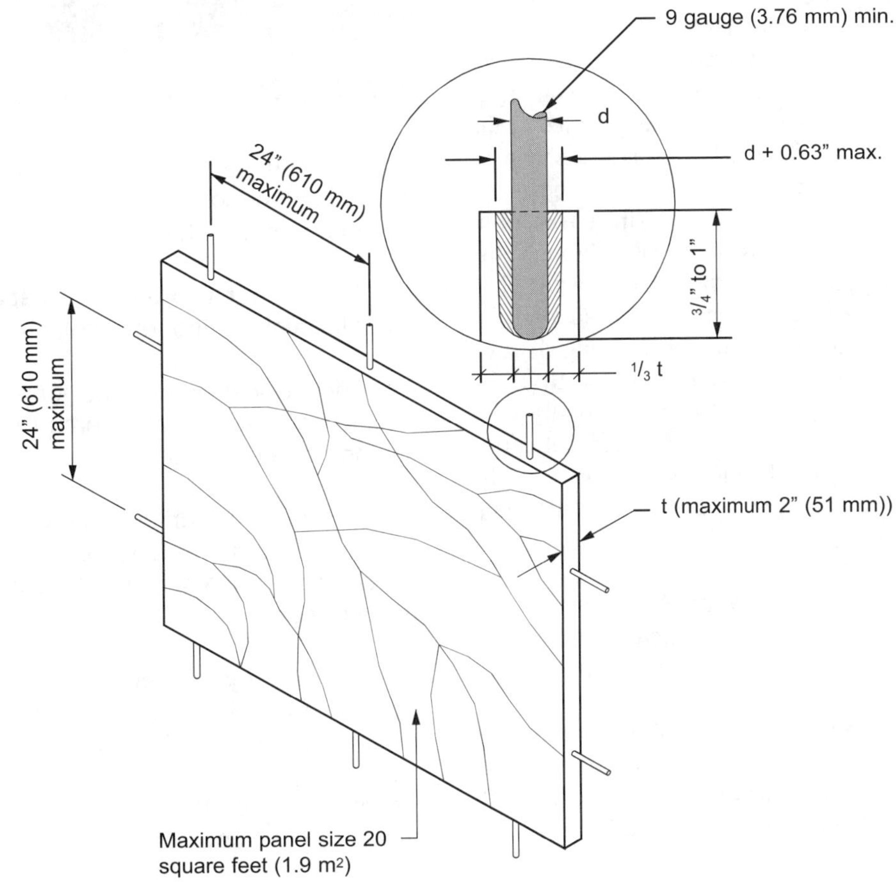

FIGURE 3.24 Anchored veneer doweling for 2 inch (51 mm) stone.

3.5 TERRA COTTA

Terra cotta, used in construction since ancient times, has enjoyed a renewed popularity in North America from the late 19th century through the 1920's as a cladding material on skeletal-framed structures. It served as an excellent cladding which architects appreciated for its fire safety, light weight and low cost, as well as for its aesthetic appeal.

Terra cotta can be molded into any design or architectural shape. It is frequently used for special ornamental work and architectural detail.

In recent years, increasing interest in the preservation and conservation of historic structures has returned terra cotta to the attention of the building industry. The material is also being used increasingly in new construction, both for cladding and ornamental purposes.

IBC Section 1405.8

1405.8 Terra cotta. Anchored terra cotta or ceramic units not less than 1.625 inches (41 mm) thick shall be anchored directly to masonry, concrete or stud construction. Tied terra cotta or ceramic veneer units shall be not less than 1.625 inches (41 mm) thick with projecting dovetail webs on the back surface spaced approximately 8 inches (203 mm) o.c. The facing shall be tied to the backing wall with corrosion-resistant metal anchors of not less than No. 8 gage wire installed at the top of each piece in horizontal bed joints not less than 12 inches (305 mm) nor more than 18 inches (457 mm) o.c.; these anchors shall be secured to 0.25-inch (6.4 mm) corrosion-resistant pencil rods that pass through the vertical aligned loop anchors in the backing wall. The veneer ties shall have sufficient strength to support the full weight of the veneer in tension. The facing shall be set with no less than a 2-inch (51 mm) space from the backing wall and the space shall be filled solidly with portland cement grout and pea gravel. Immediately prior to setting, the backing wall and the facing shall be drenched with clean water and shall be distinctly damp when the grout is poured.

Unlike adhered terra cotta, anchored terra cotta is not limited in unit size or area, except as required for the control of expansion and contraction.

Facing of the anchored terra cotta must be tied with at least 8 gauge (4.1 mm) noncorrosive wire placed in horizontal bed joints and vertical head joints between 12 in. (305 mm) and 18 in. (457 mm).

FIGURE 3.25 Terra cotta cornice.

These anchors are to be tied to $^1/_4$ in. (6.4 mm) galvanized pencil rods vertically aligned with loop anchors that are attached to the backing wall.

The veneer ties must be capable of supporting the full weight of the terra cotta veneer in tension.

Prior to installation, the terra cotta and backing should be soaked in the same manner as described for the adhered terra cotta.

A solid fill of portland cement grout between the backing and the anchored terra cotta should be accomplished maintaining at least a 2 in. (51 mm) grout space.

As with any veneer system, consideration must be given for moisture penetration. Even though there is little water penetration through the face of the terra cotta unit, the system is still subject to moisture penetration, so the veneer system should include flashing and weep holes.

Manufacturers of terra cotta can be extremely helpful in suggesting attachments that conform to the Code and are practical to use.

In commercial applications, the manufacturer will become integrally involved with the project, and is a reliable source of shop drawings. Since there can be many unique pieces in the terra cotta project, it is not unusual for the supplier to detail each and every piece, assigning an individual sequence number for each shape.

Most manufacturers also possess the technical expertise to analyze, develop and test terra cotta so that the existing terra cotta may be matched or improved.

3.6 BLOCK VENEER, CONCRETE UNITS

Block veneer can be processed with quality control standards that allow color and texture variations similar to clay brick. Since the manufacturing process does not require the high firing temperatures associated with clay products, block veneer is cost competitive.

Block veneer can also be made into a variety of textures, such as smooth, slumped or adobe faced, split faced and fluted.

Quite often, however, the application of block veneer may be impractical. Alternatively, a wider structural block may be used to achieve both a desired surface appearance and a structural element potentially yielding a substantial cost savings.

Block veneer may not be as readily available as clay brick veneer units, especially in the smaller, more traditional, clay brick sizes.

As with any product, communication between the design team and the developer is the key in selecting the most appropriate product to satisfy the end user, while maintaining an efficient cost approach.

FIGURE 3.26 Anchored block veneer.

FIGURE 3.27 Anchored block veneer system with concrete backup.

4 TYPICAL DETAILS

4.1 GENERAL

The following details are presented as a few of the design possibilities of a veneer system.

4.2 ROOF/PARAPET DETAILS

Each project should be designed on its own merit using details most beneficial for the application.

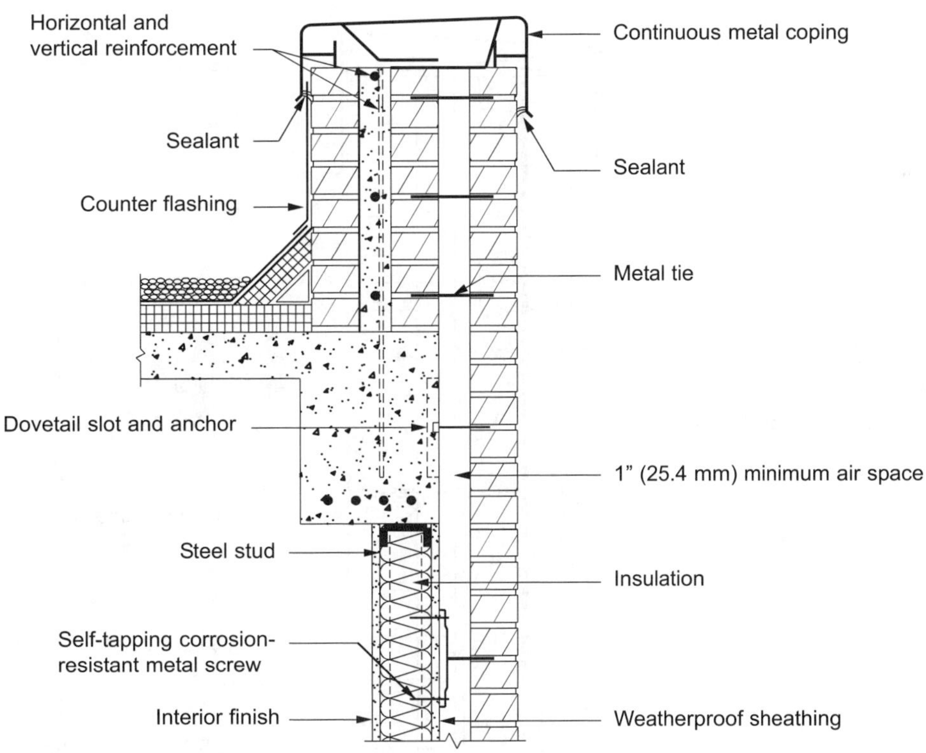

FIGURE 4.1 Brick parapet on concrete frame building.

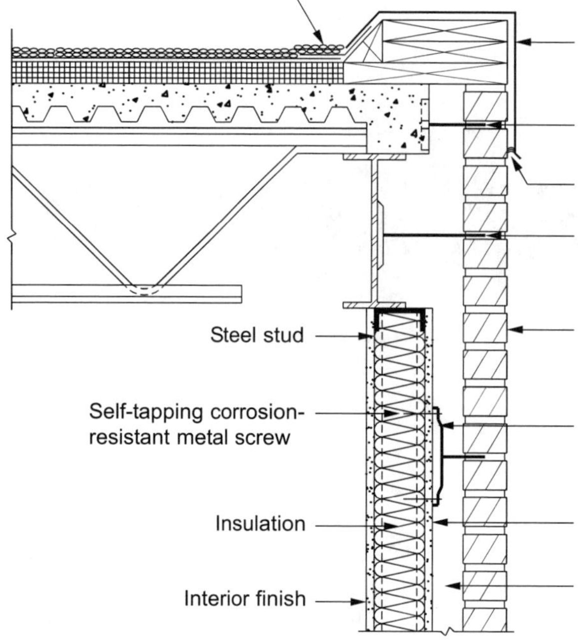

FIGURE 4.2 **Roof detail without parapet.**

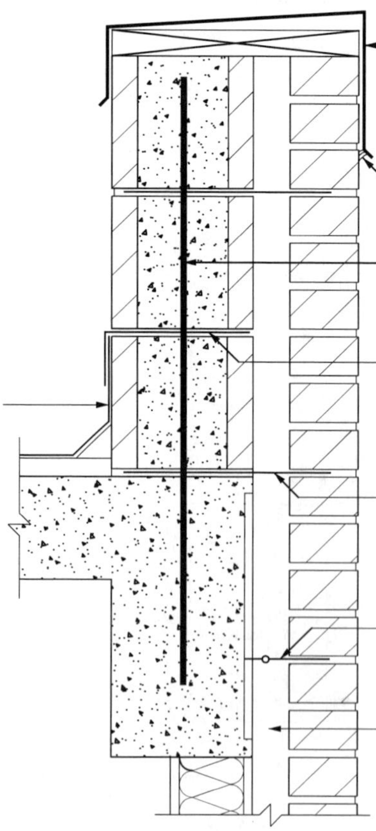

FIGURE 4.3 **Masonry parapet wall.**

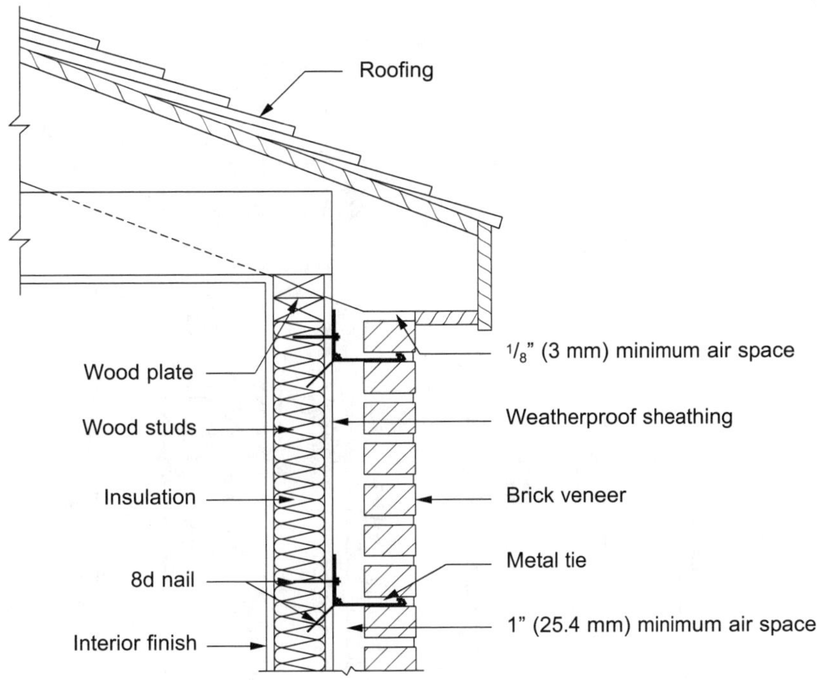

FIGURE 4.4 Sloped wood roof detail.

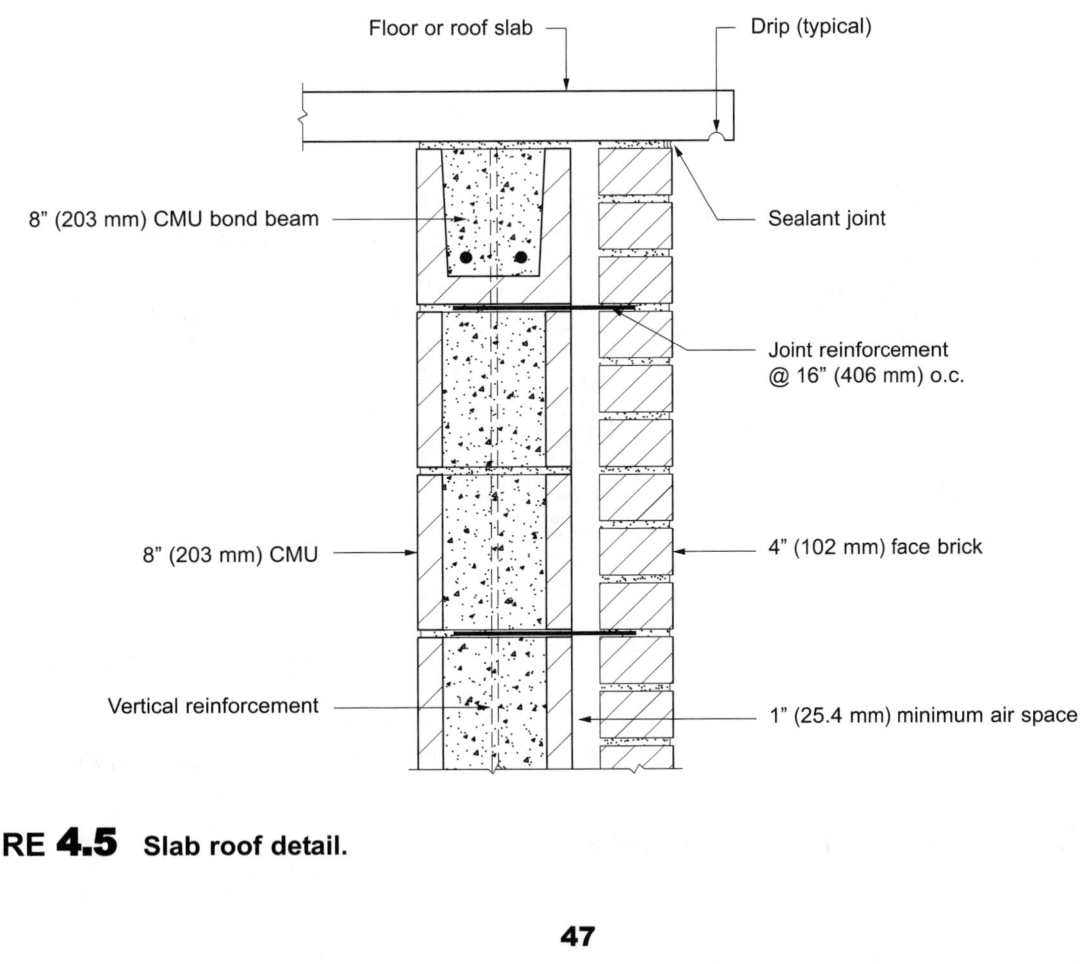

FIGURE 4.5 Slab roof detail.

4.3 SHELF ANGLE/FLASHING DETAILS

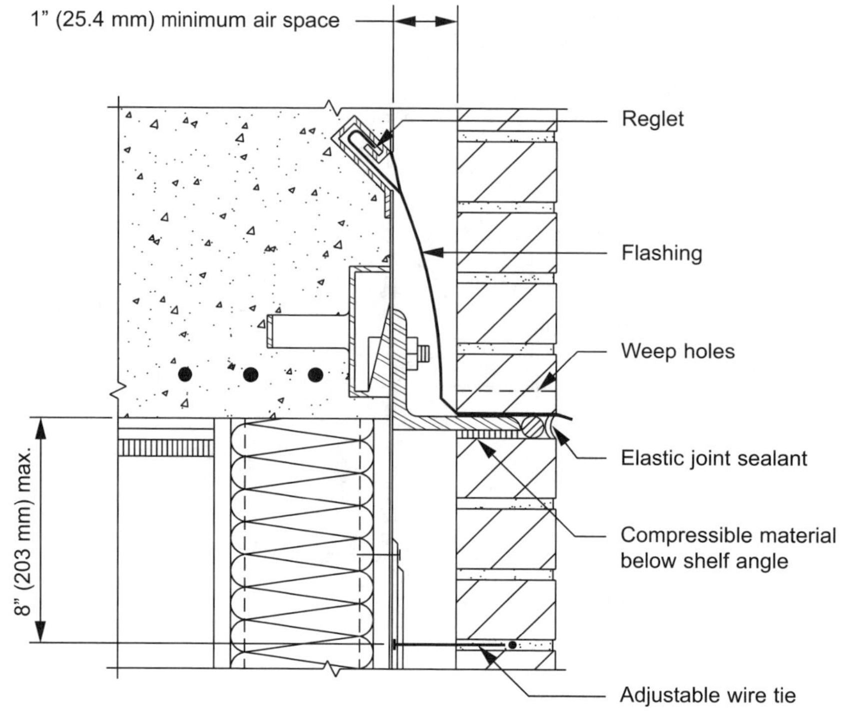

FIGURE 4.6 Flashing at shelf angle.

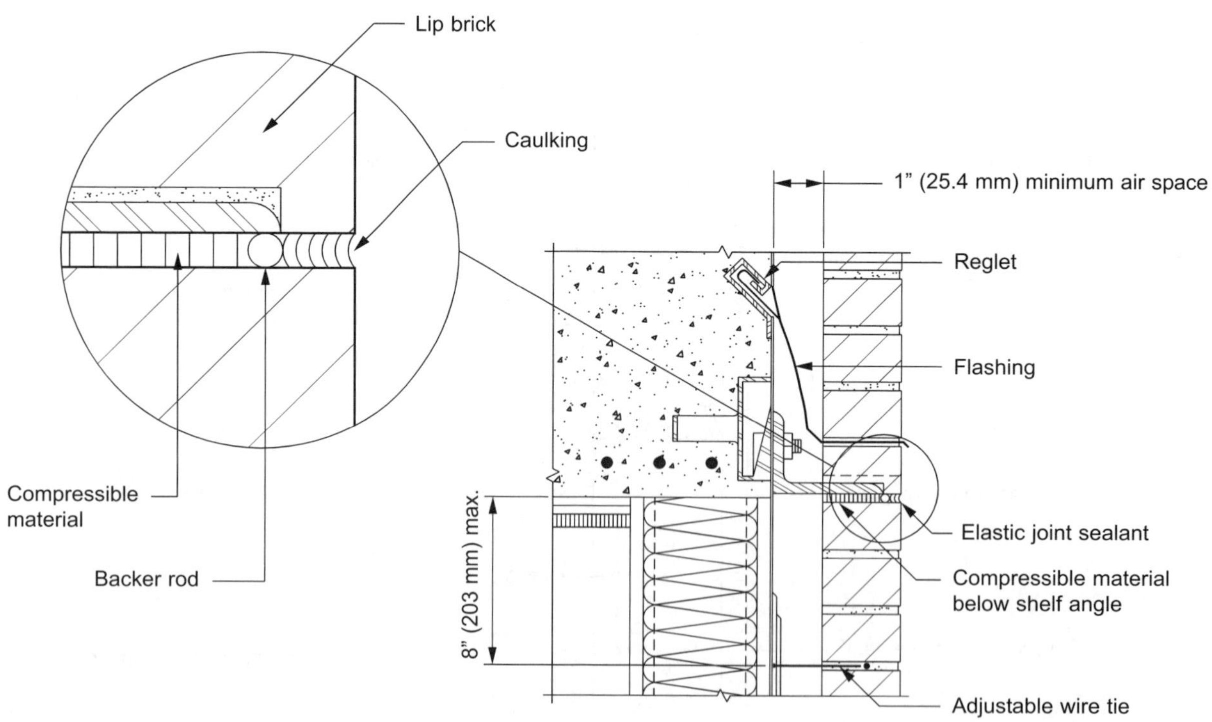

FIGURE 4.7 Flashing one brick above shelf angle.

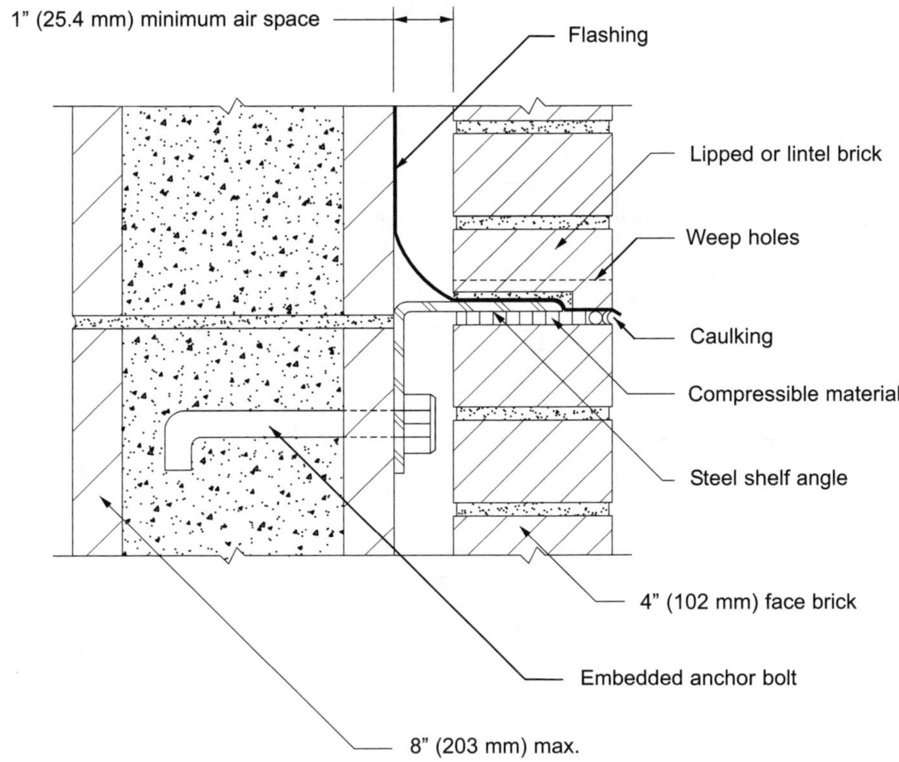

FIGURE 4.8 **Shelf angle anchored to CMU backup.**

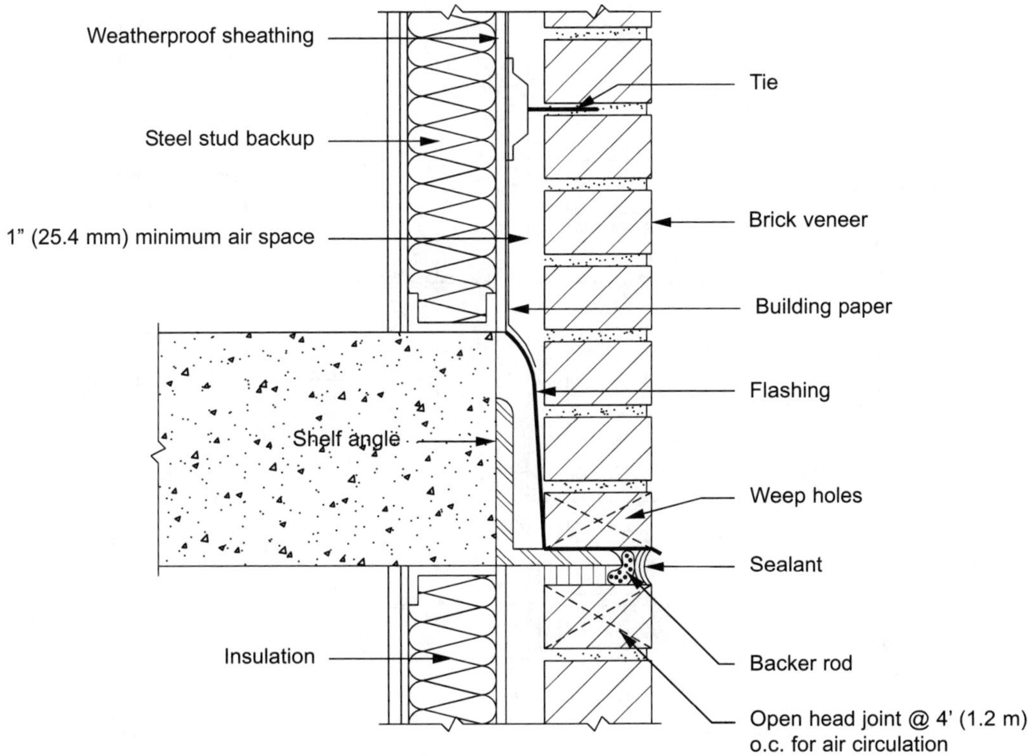

FIGURE 4.9 **Shelf angle at concrete support, steel stud backup.**

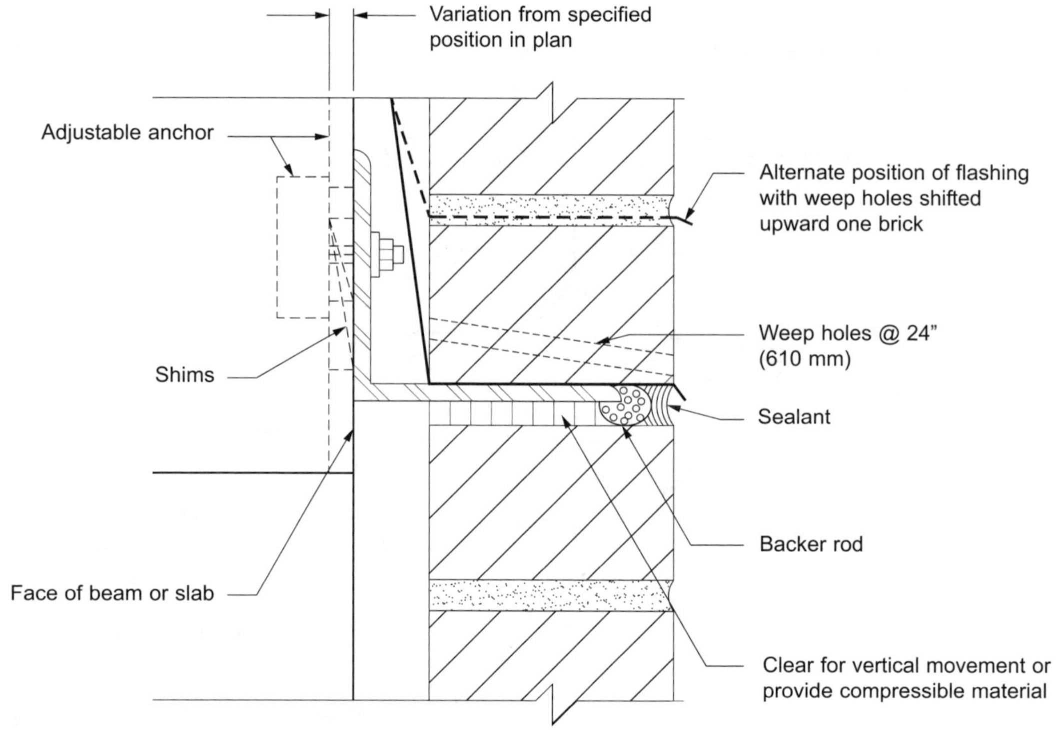

FIGURE 4.10 **Shelf angle anchored to beam or slab.**

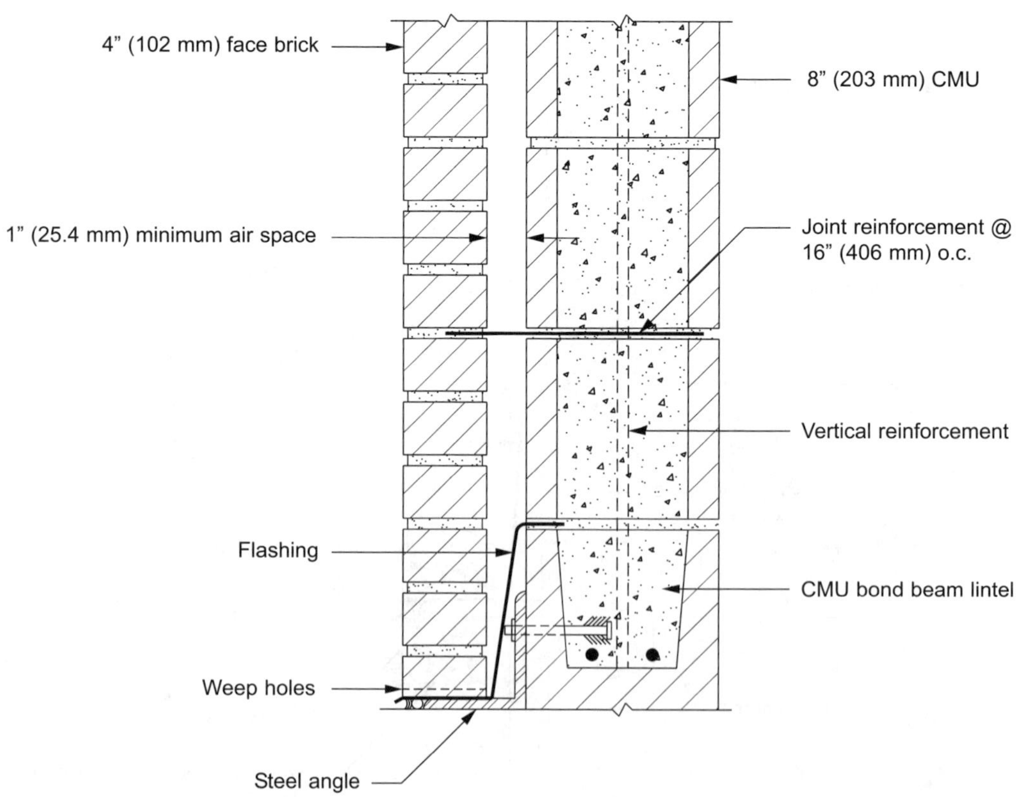

FIGURE 4.11 **Shelf angle anchored to CMU backup.**

4.4 SILL AND JAMB DETAILS

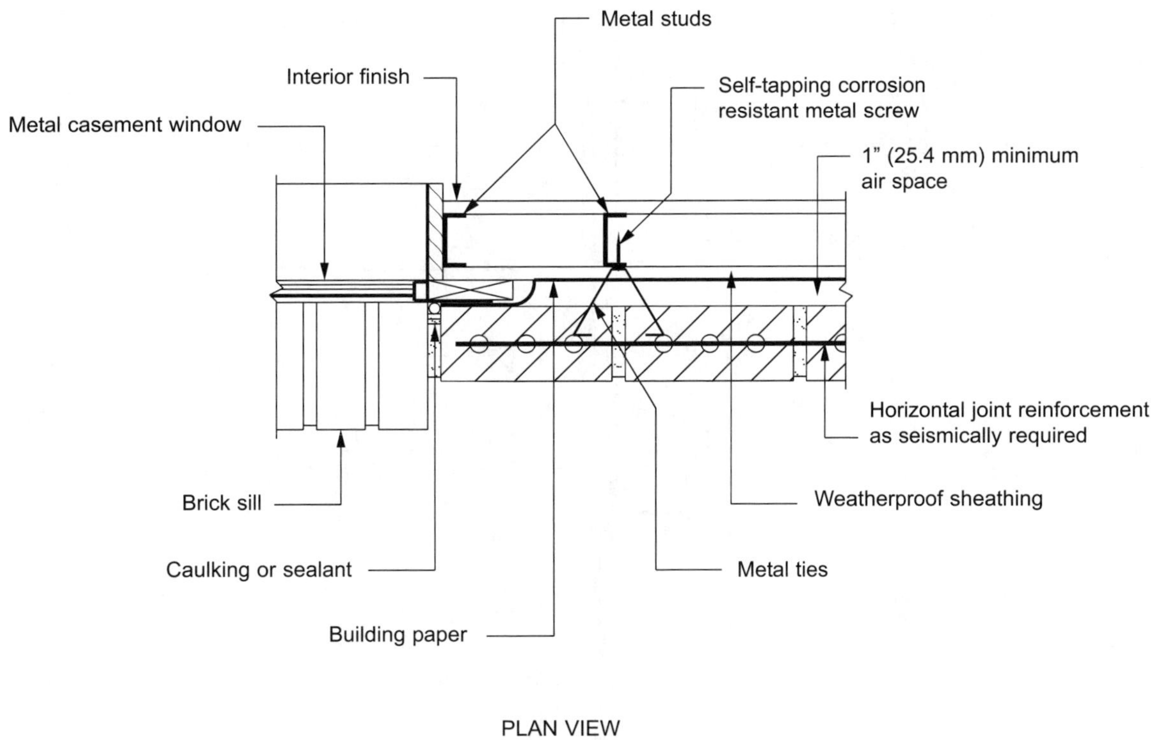

FIGURE 4.12 Window jamb, steel studs/brick veneer.

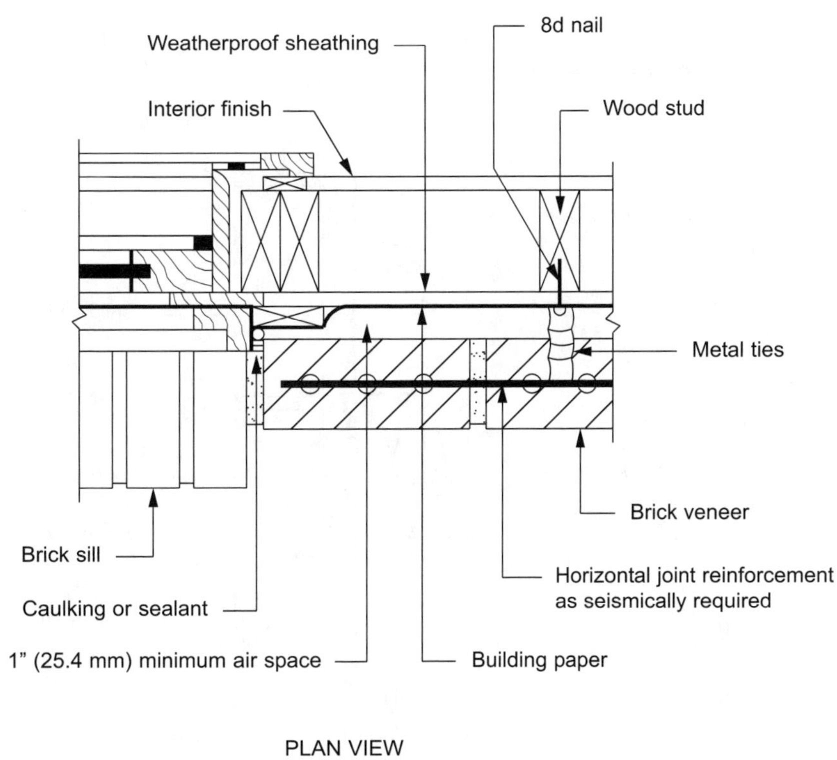

FIGURE 4.13 Window jamb, wood studs/brick veneer.

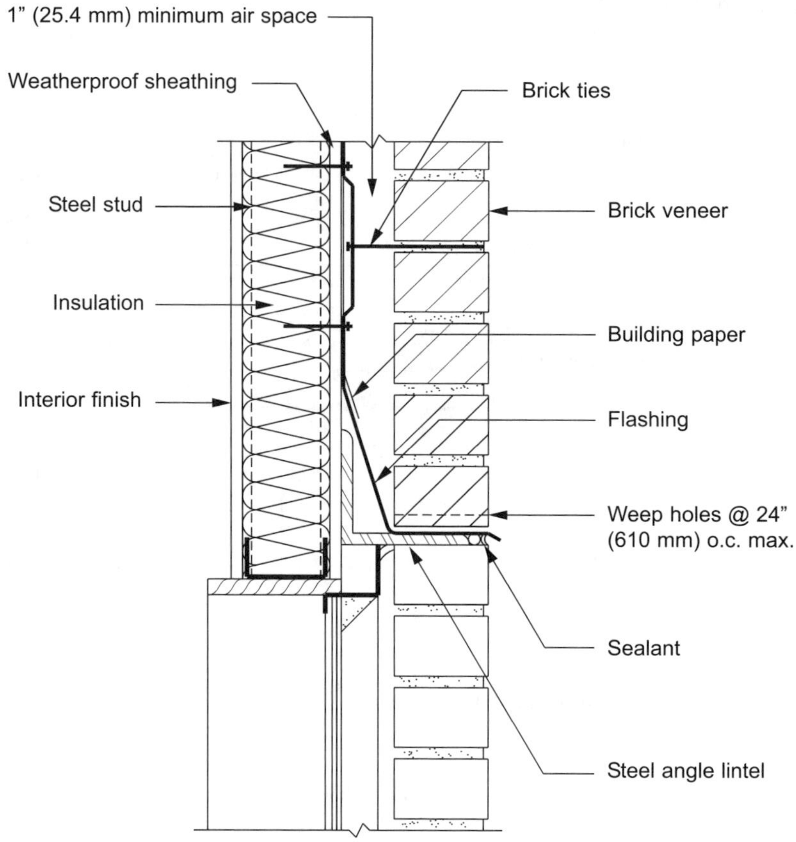

FIGURE 4.14 Window/door soffit steel studs/brick veneer.

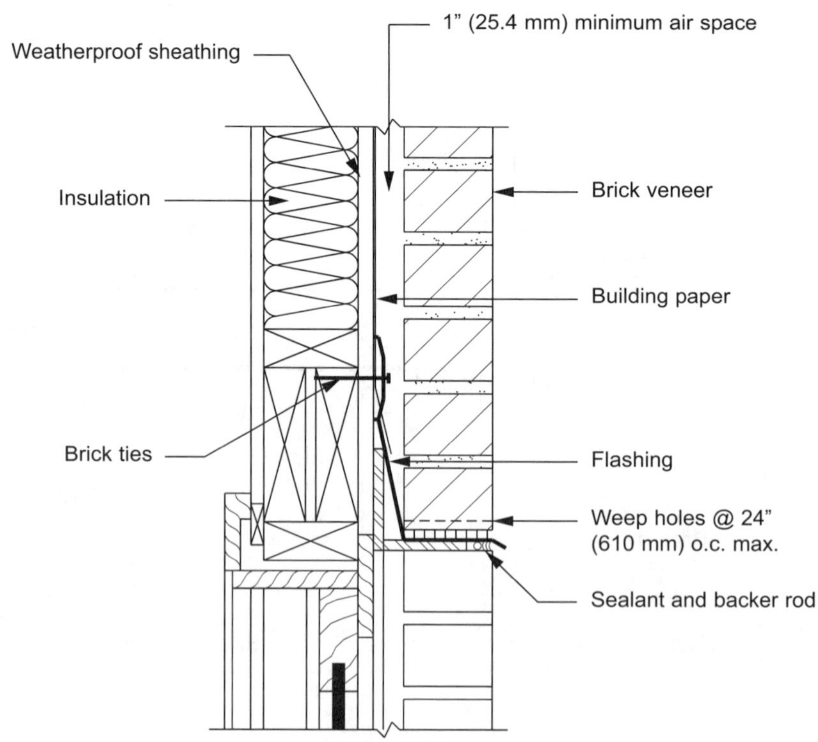

FIGURE 4.15 Window/door soffit wood studs/brick veneer.

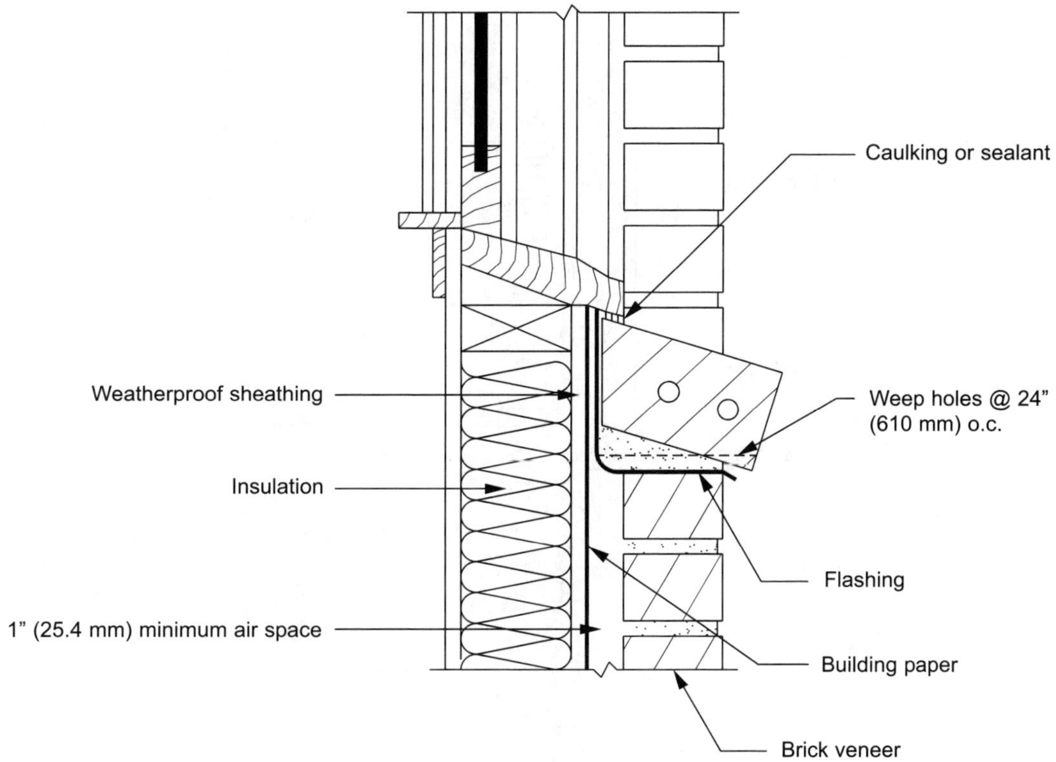

FIGURE 4.16 **Window sill, wood studs/brick veneer.**

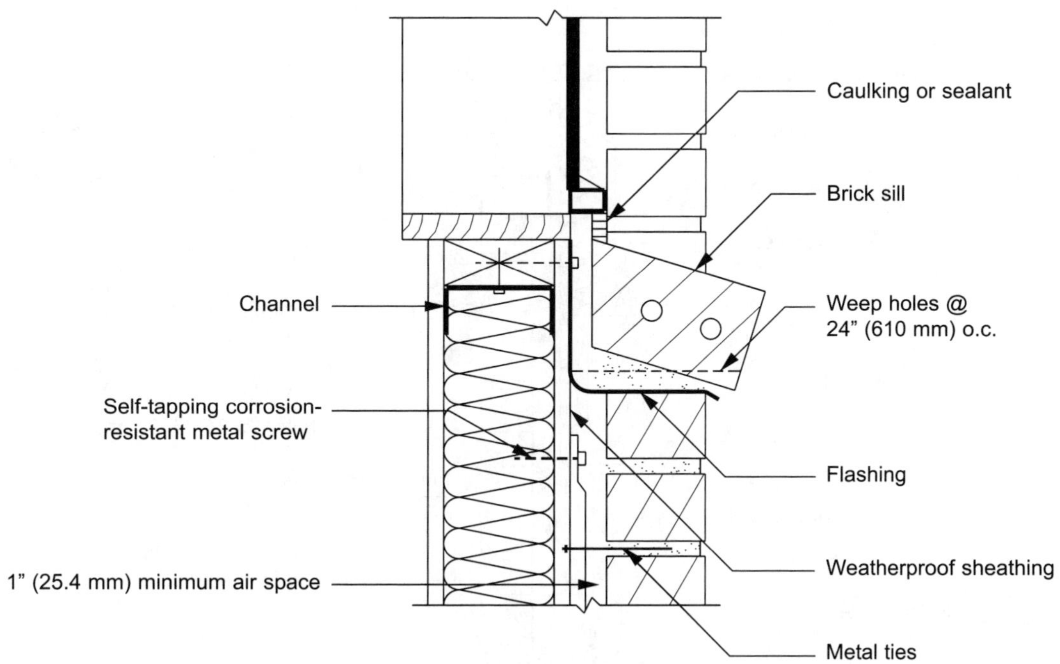

FIGURE 4.17 **Window sill, steel studs/brick veneer.**

4.5 FLOOR CONNECTION DETAILS

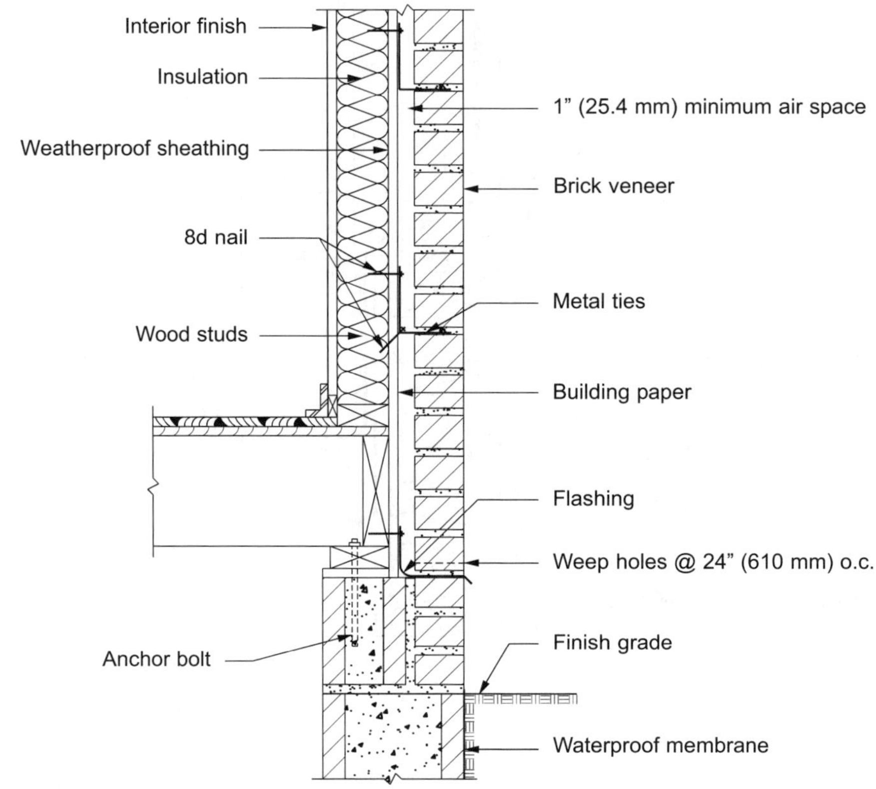

FIGURE 4.18 Floor connection detail.

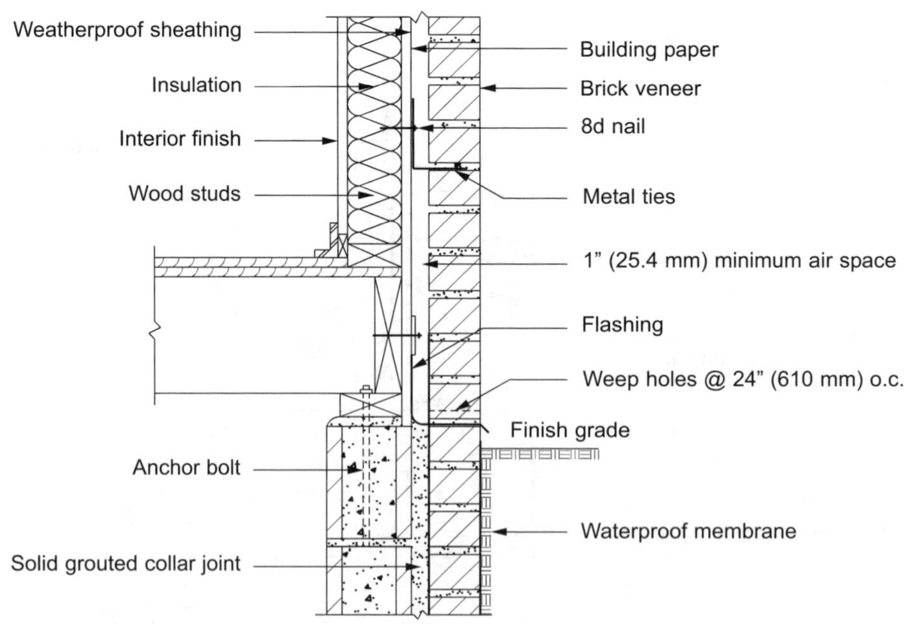

FIGURE 4.19 Floor connection detail.

4.6 WALL BASE DETAILS

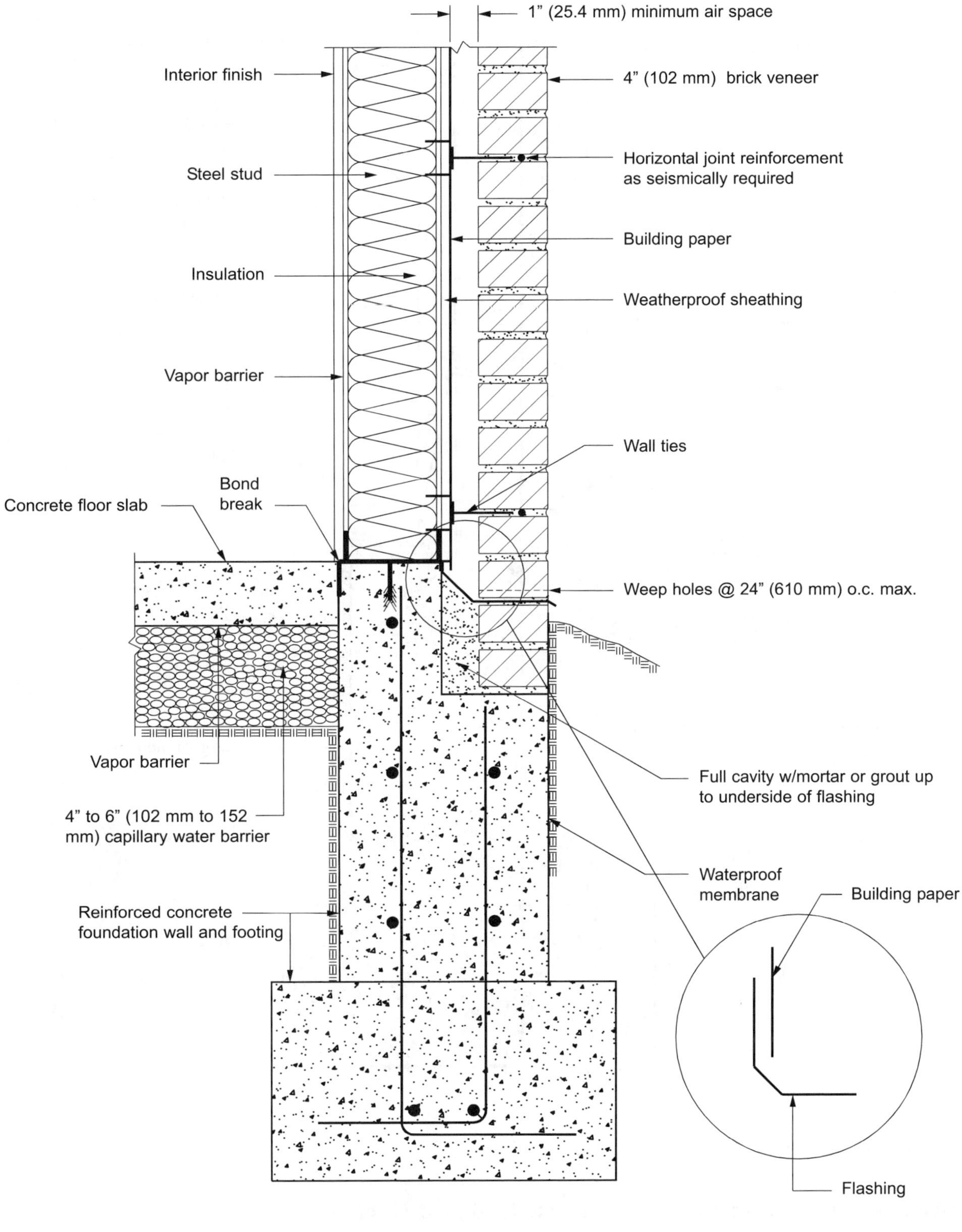

FIGURE 4.20 Base of wall detail, concrete footing, steel stud backup.

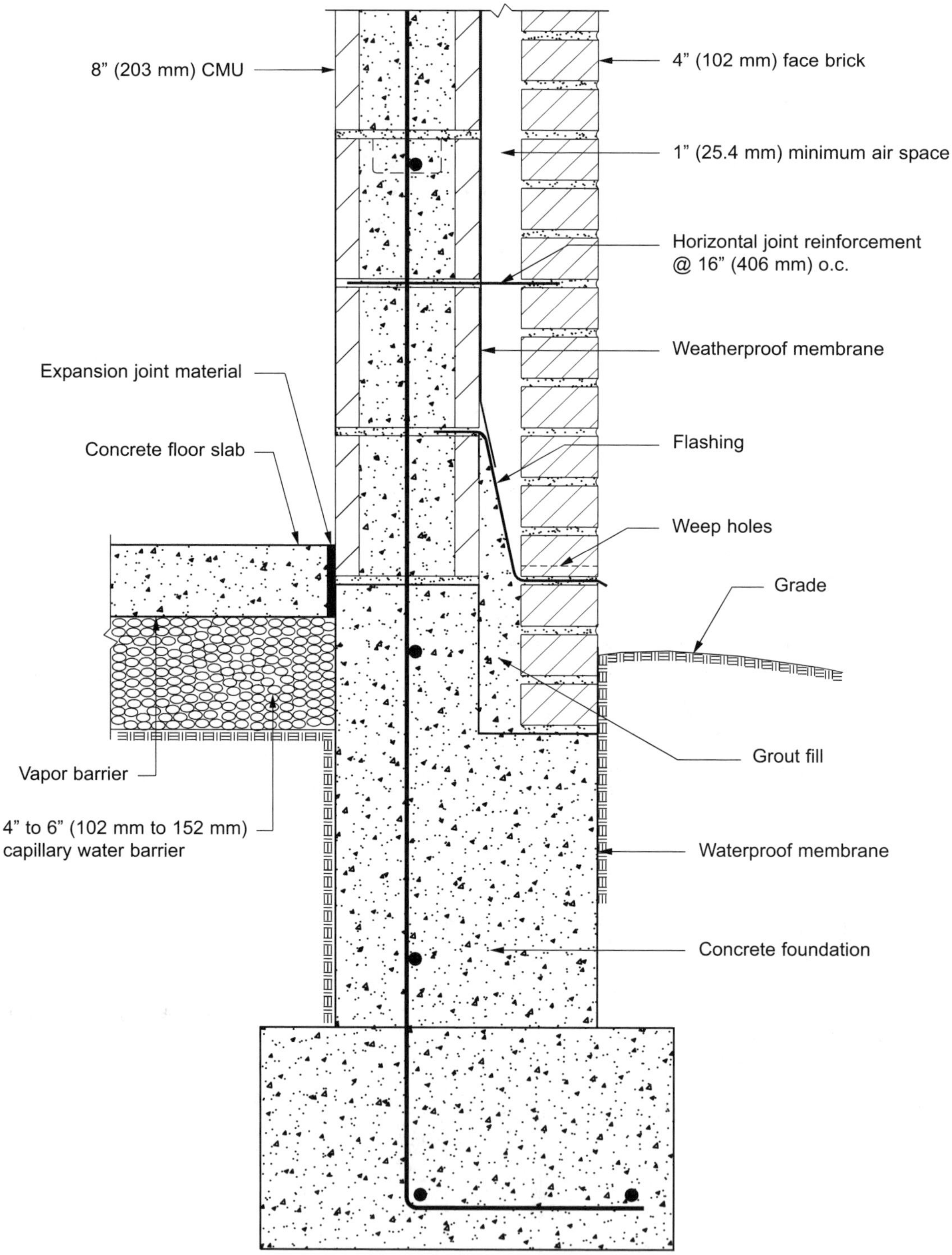

FIGURE 4.21 Base of wall detail, concrete footing, CMU backup.

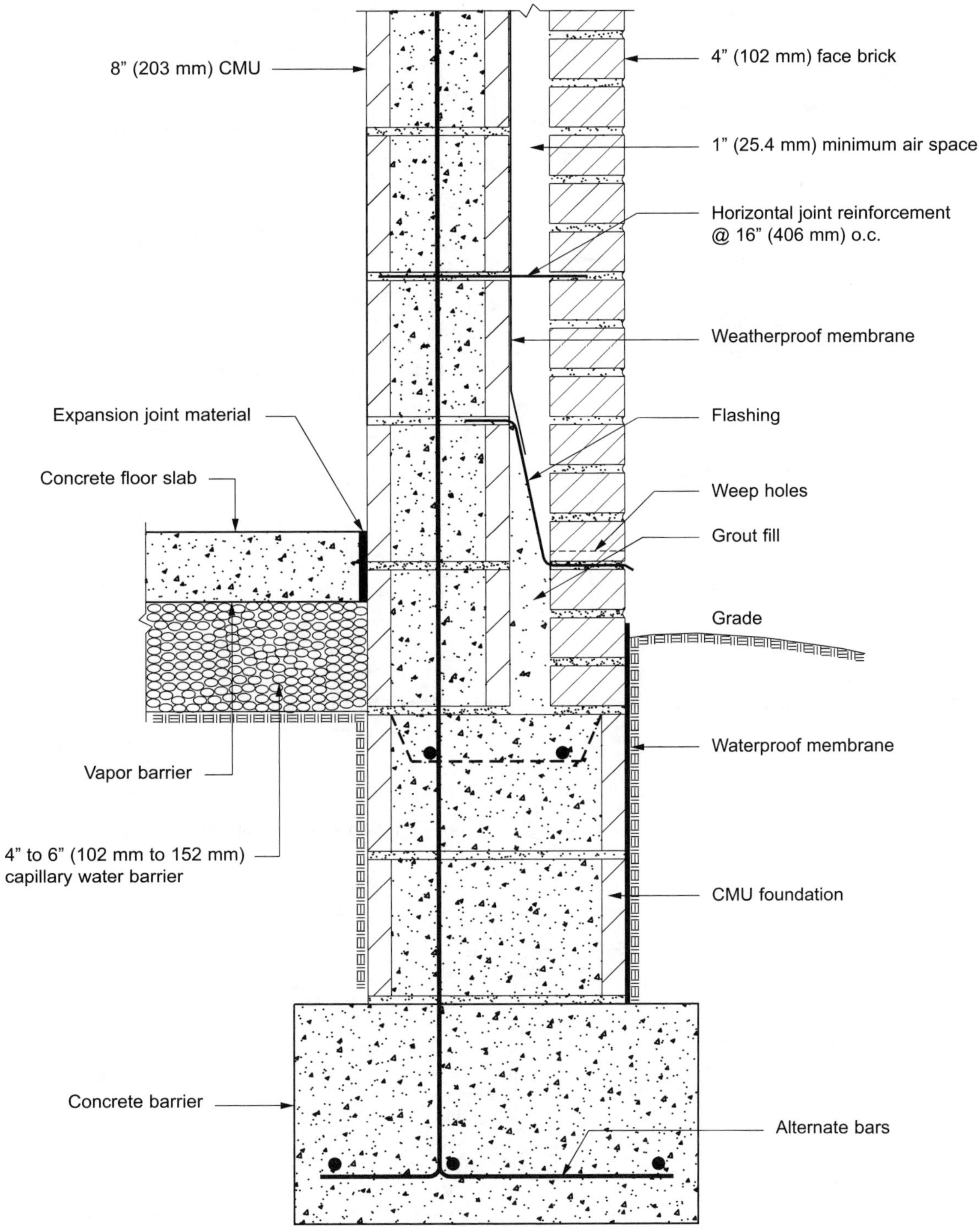

FIGURE 4.22 **Base of wall detail, concrete footing CMU below grade, CMU backup.**

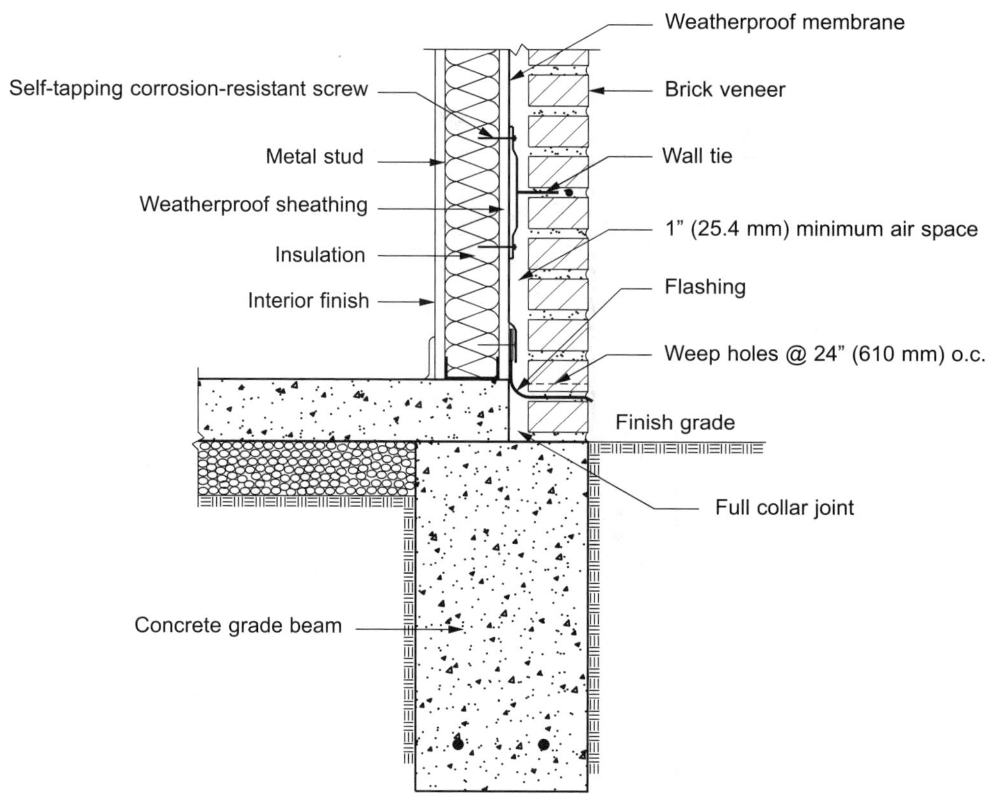

FIGURE 4.23 **Base of wall detail, concrete footing, steel stud backup.**

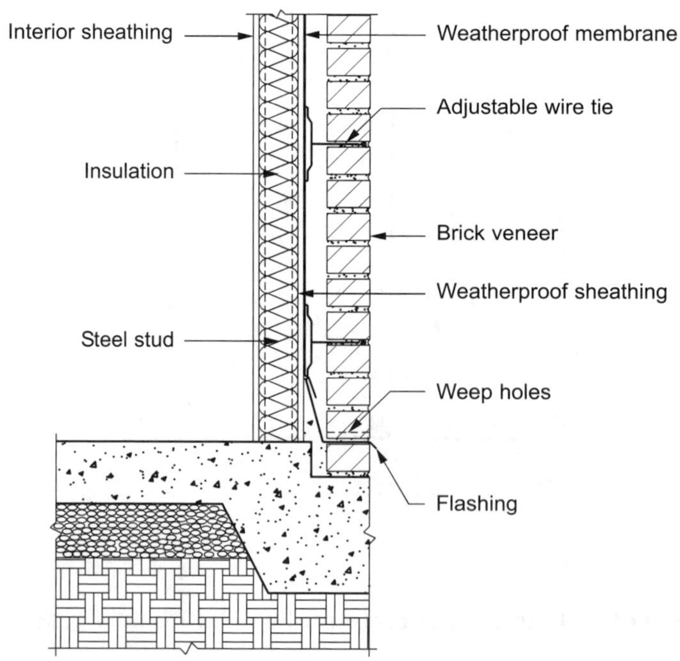

FIGURE 4.24 **Base of wall detail, concrete footing/slab, steel stud backup.**

4.7 EXPANSION JOINT DETAILS

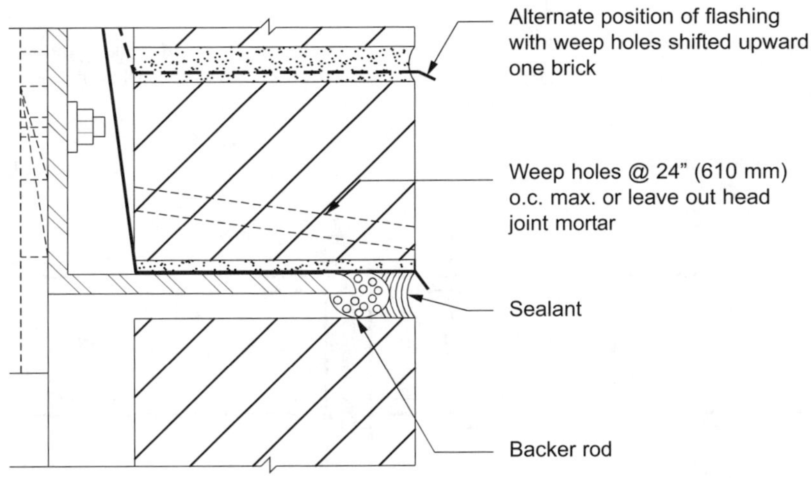

FIGURE 4.25 Horizontal expansion joint at shelf angle.

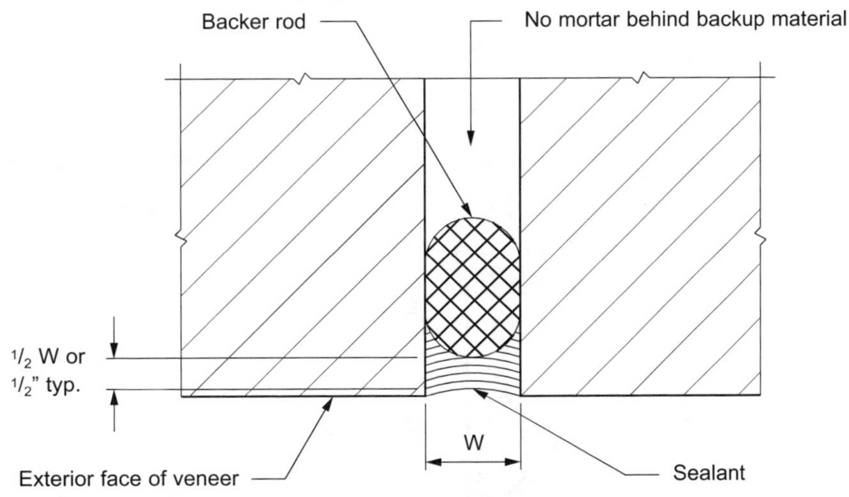

FIGURE 4.26 Plan view of vertical expansion joint.

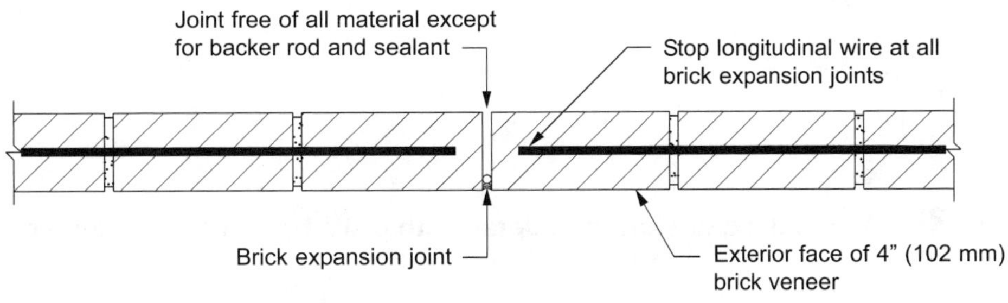

FIGURE 4.27 Plan view of horizontal reinforcement expansion joint.

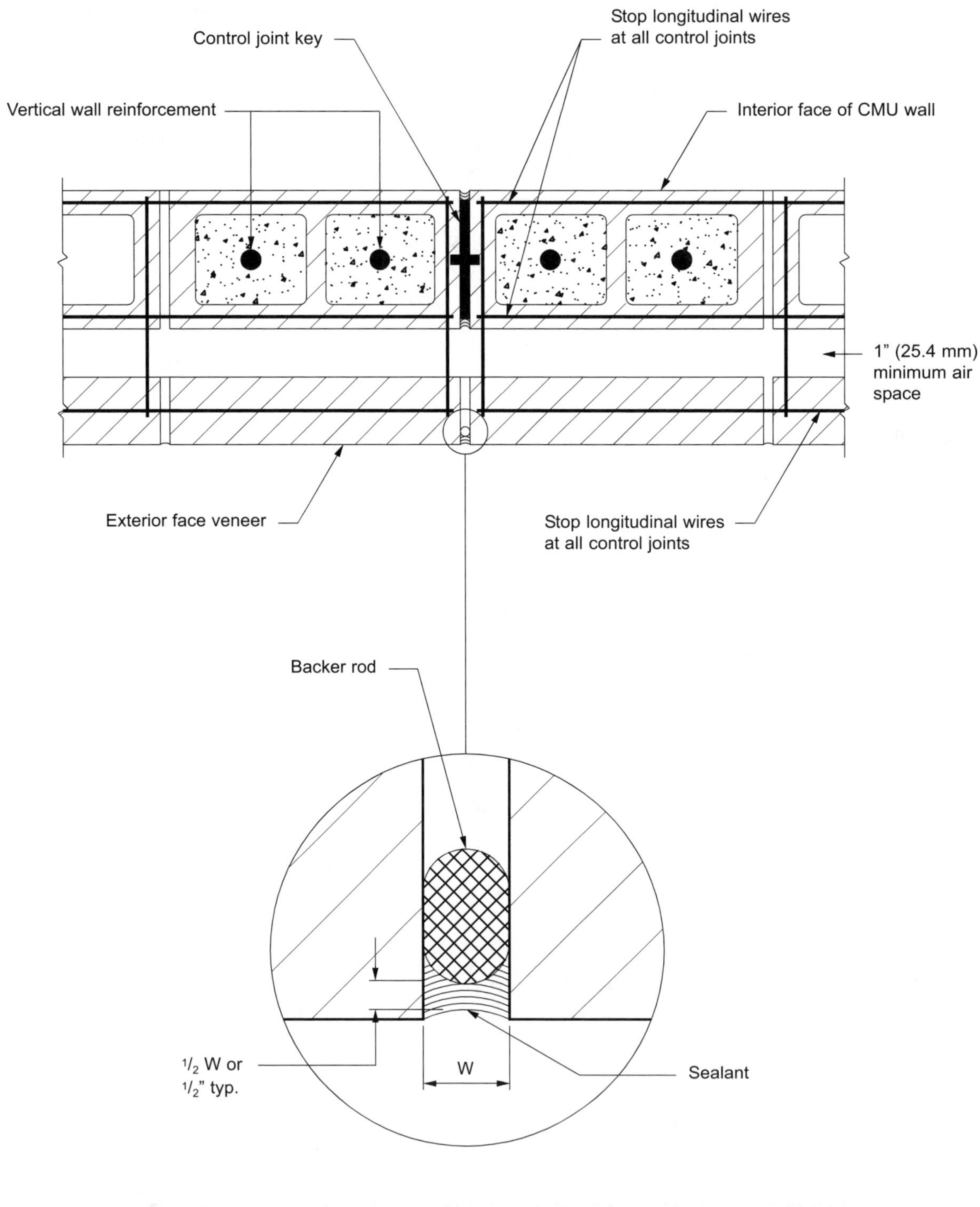

FIGURE 4.28 **Vertical expansion joint detail with CMU backup/brick veneer.**

4.8 SYSTEM CONFIGURATION

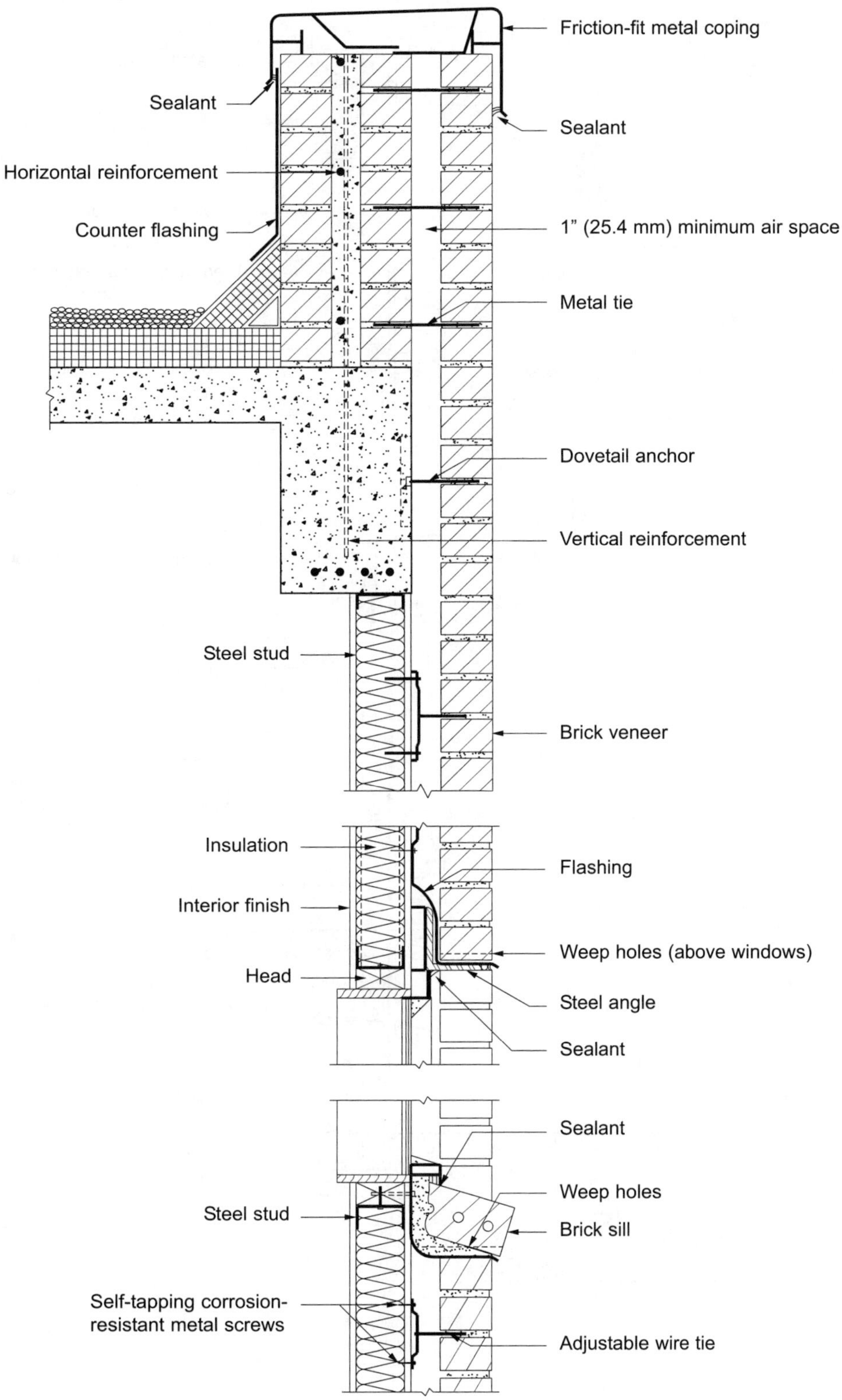

FIGURE 4.29 Veneer system.

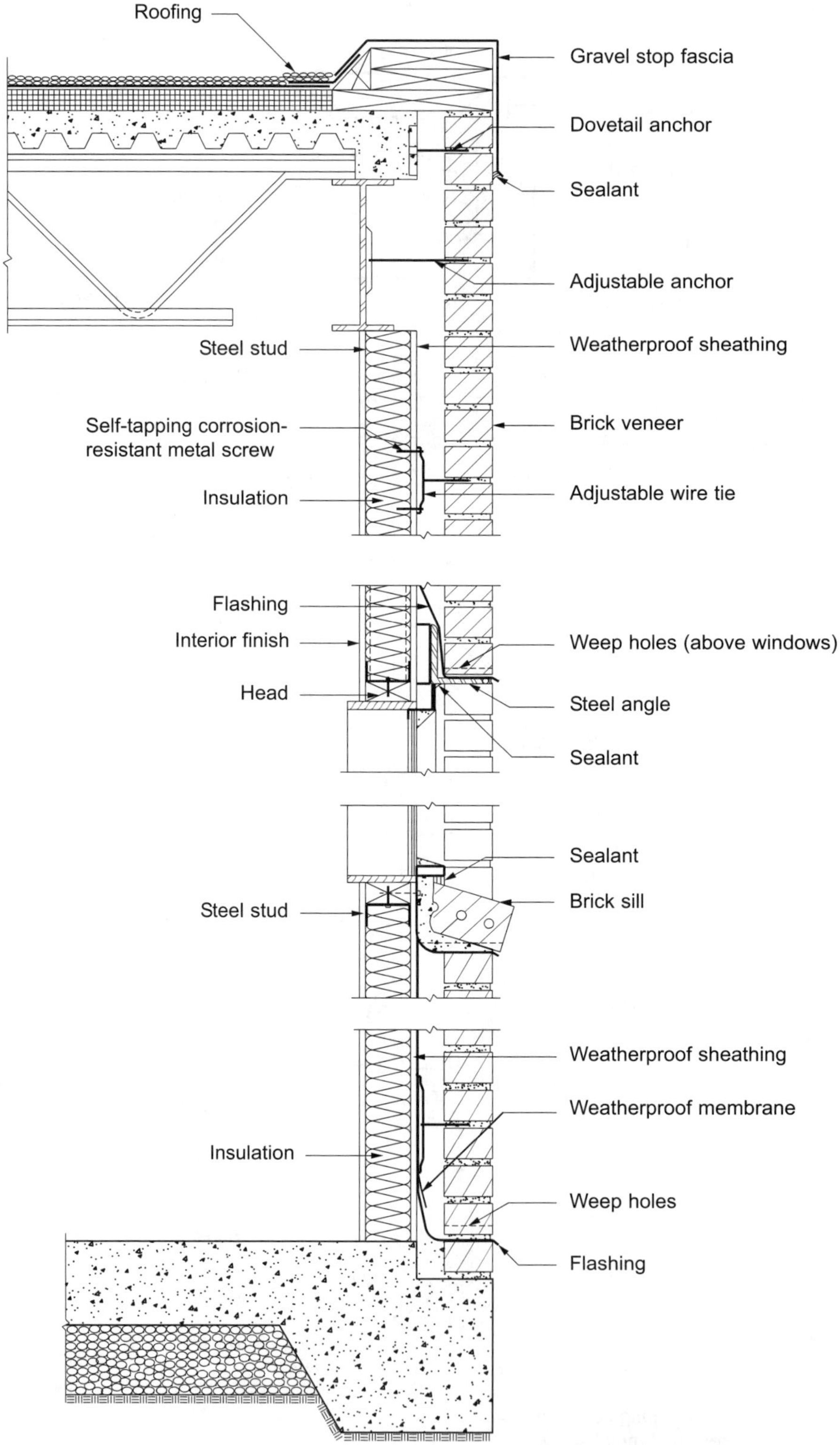

FIGURE 4.30 **Veneer system.**

5 GUIDE SPECIFICATIONS

5.1 GUIDE SPECIFICATION FOR ADHERED THIN CLAY BRICK OR STONE SLAB MASONRY VENEER

GUIDE SPECIFICATION	NOTES TO SPECIFIER
PART 1- GENERAL	
1.01 Division 1 Requirements, Drawings, General and Supplementary General Conditions, and Special Conditions apply to this section.	**1.01** This includes all the work the masonry contractor is to accomplish.
1.02 WORK INCLUDED Furnish and install brick or stone slab veneer units and mortar. Provide equipment necessary for their installation.	**1.02** and **1.03 "WORK INCLUDED"** and **"RELATED WORK"** are listed here to define these facets, though they are sometimes not made part of the Division Specifications. If these paragraphs are included, the Specifier must check that there are not portions overlapping or not covered elsewhere.
1.03 RELATED WORK SPECIFIED ELSEWHERE	
A. Plaster undercoats: Section 09____.	**1.03.A.** Plaster undercoats may be part of "lath and plaster" or part of this Section.
B. Rough Carpentry: Section 06____.	
C. Caulking, Sealants and Waterproofing: Section 07____.	
1.04 QUALITY ASSURANCE	**1.04** Specify any special submittal or tests required to assure the materials and construction procedure will be of the quality desired.

A. Submittals:

1. Sample: Submit ____ brick or stone slab veneer units of each type, including special shape required to show range of colors, textures, finishes and dimensions.

2. Certification: Furnish Manufacturer's certification that the brick units provided meet or exceed the requirements of this specification.

1.04.A Specify the number of veneer units required for approval. Coordinate submittal requirements with standard procedures.

B. Sample Panels:

1.04.B Sample panels and mock-ups are a good way to assure that workmen know what is expected for them to provide. Sample panels can be made as part of the finished project.

Sample panels may not be necessary for small jobs.

1. Construct a sample panel of the work of the Section at the location on the site indicated by the Architect.

2. Provide one sample panel for each combination of brick veneer patterns, bond pattern, and mortar color.

3. Sample panels shall be ____ft (____m) x ____ft (____m).

1.04.B.3 Specify the size of sample panel required along with any special requirements on its construction.

4. Show color range, texture range, mortar color, joint tooling, bond pattern, cleaning, quality of workmanship and other aspects of the work relating to this Section.

5. Make necessary adjustments in the sample panels and secure the Architect's approval. Construction may not proceed until the Architect approves sample panel mock-up.

6. Sample panels shall remain on the project site to serve as a datum for comparison with the remainder of work of this Section for the purpose of acceptance or rejection.

7. If the sample panel is not included in the actual masonry work, it shall be demolished after the masonry work for the project has been accepted.

1.05 PRODUCT DELIVERY, STORAGE AND HANDLING

1.05 The intent of this section is to keep materials clean, dry and in an acceptable condition during and after job delivery.

A. Store brick or stone slab veneer, cement, lime and other materials of this section off the ground to prevent contamination with mud, water or other harmful materials. Veneer units shall be handled and stored in such a manner so as to prevent chipping, breaking and cracking.

B. Cover and protect materials from inclement weather to maintain quality control and physical requirements.

PART 2- PRODUCTS

2.01 BRICK VENEER UNITS

2.01 Numerous different types, sizes, styles and colors of veneer units are available. Verify that the color, texture and dimensions selected are available from the Specified Manufacturer.

For thin brick units, specify the grade (exterior or interior) and type (TBX, TBS or TBA). Type TBS (Standard) thin brick is for general use in exposed exterior and interior masonry walls. For a higher degree of precision and a lower variation in size, specify Type TBX (Select) units. Type TBA (Architectural) brick veneer units have intentional nonuniformities in size, color and texture.

A. Units shall have ______face texture and ______ color as manufactured by __________.

B. Reference Specification: ASTM C1088 –____, *Standard Specification for Thin Veneer Brick Units Made from Clay or Shale,* Type ____, Grade ____.

C. Dimensions: ____ in. (____ mm) wide by ____in. (____ mm) long by ____ in. (____ mm) thick.

D. Provide all units with straight cut edges and square corners unless noted otherwise.

E. Do not permit brick veneer units to vary more than $^1/_8$ in. (3 mm) in any dimension from those specified. Provide units similar in texture and color to those approved by the Architect.

2.02 STONE SLAB UNITS:

2.02 Specify the types, sizes, grades and colors of the desired veneer units. Add any special requirements as necessary.

If a travertine stone is selected, indicate whether it is filled or unfilled.

Surface finishes vary considerably between different stone materials; specify accordingly. When two or more surface finishes are required, consider using a schedule at the end of this Section.

A. Limestone: [Cut Indiana Oolitic limestone.] __________.

B. Marble: ASTM C503, Classification [I___Calcite;] [II___Dolomite;] [III___Serpentine;] [IV__Travertine;] [with [filled] [unfilled] surface].

C. Granite: ASTM C615, [sawed] [cut] [split] ______.

D. Slate: ASTM C629, Classification [I___Exterior;] [II___Interior;] [sawed] [cut] [split] __________.

E. Sandstone: ASTM C616 Classification [I___Sandstone] [II___Quartzitic Sandstone] [III___Quartzite].

F. Color: __________.

G. Surface Finish: __________.

H. Grade:__________; free of defect.

I. Thickness:______in. (____mm) nominal.

J. Face Size:____x____in. (___mm x___mm) nominal.

2.03 MORTAR:

2.03 Mortar materials may be specified in a separate specification or in this specification. The editor's have selected to provide a separate mortar specification for simplicity in this publication. See Guide Specification for Mortar (Section 04100).

A. Mortar shall comply with the requirements of Section 04100 with the following exceptions:

1. Thin set mortar (if used): Provide a commercial grade mortar specifically formulated for the purpose and composed of latex mixed with portland cement, or provide such other thin set mortar as is specifically approved in advance by the Architect.

2.03.A.1 Commercially prepared latex mortars are available and have been used with great success. Consult manufacturer for detailed information and help in selection.

2. Pointing mortar: Provide a mix of one part portland cement and ______ part lime to ______ parts sand by volume.
Color shall be __________.

2.03.A.2 Pointing mortar, if not incorporated with the setting process, should be specified and should state color desired. Typically pointing mortar is type N mortar with 1 part portland cement, $^{1}/_{2}$ to $1^{1}/_{4}$ parts lime and sand between $2^{1}/_{4}$ and 3 times the volume of cementitious material.

PART 3 - EXECUTION

3.01 SURFACE PREPARATION

A. Examine the areas and conditions under which work of this Section will be performed. Correct conditions detrimental to timely and proper completion of the work. Do not proceed until unsatisfactory conditions are corrected.

3.01.A The substrata must be prepared properly so that the veneer will perform properly. The supporting surface should be relatively rigid so that it may support the thin veneer adequately. The presence of cracks or yielding of the surface should be checked as they may indicate an unsatisfactory surface.

B. Thoroughly clean the substrata of debris, dust, and finishes which will not bond with the mortar.

1. Remove sealers, bond breakers, and other applied finishes by use of a light sandblast or other means approved by Architect.

3.01.B.1 Proper bond depends upon having suitable substrata. The substrata must be compatible with portland cement and must be clean. Oil, dust residues, sealers, bond breakers, form releases and other substances can significantly reduce the bond adhesion between the mortar and the substrata and therefore must be completely removed. Sandblasting is one of the easiest and best methods to remove these substances although chipping, scraping and wire brushing have also been used successfully.

2. Provide substrata with the ability to receive and bond with the mortar. Dampen the backing surface with clean water by fogging. Surface shall be damp, but not wet or saturated, at the time of laying.

3.01.B.2 The area to be faced is often fog sprayed about 2 hours prior to veneer mortar installation. This procedure keeps the water in the mortar from being sucked into the substrata too quickly which can reduce the bond adhesion. The actual time and amount of wetting depends on the absorption of the backing materials, the temperature and the humidity. Note that extremely absorptive veneer units are sometimes wetted prior to laying as well.

3.02 INSTALLATION

A. General

1. Position all veneer units level, plumb and true and in uniform coursing in accordance with the Drawings. All corners and angles shall be square unless otherwise indicated on the drawings.

3.02.A.1 Tolerances for manufacturing and installation must be considered. An appropriate installation tolerance guide is contained in Specification for Masonry Structures (ACI 530.1-05/ASCE 6-05/TMS 602-05), Article 3.3 G.

2. Unless otherwise indicated on the Drawings, place units in running bond. Install control and expansion joints as shown on the drawings and at all control joints in the backing surface.

3.02.A.2 State the desired bond pattern. Place expansion and control joints as required by the materials and as directed by the building code. Joints in the veneer should always be located at joints in the substrata to allow adequate shrinkage and temperature movements and to avoid cracking of the veneer.

3. Lay out each wall or panel in a manner to minimize cutting of brick veneer units. Use masonry saws to cut and fit masonry units.

B. Spread the approved mortar on the surface of the setting bed and the back of the veneer units using a notched trowel. Provide a uniform setting bed of $^1/_8$ in. minimum thickness.

3.02.B In some instances it may be advantageous to permit the mason to only apply the mortar to the back of the veneer surface. By fully covering the back surface of the veneer only, the mason can lay the units much more rapidly and align the units with wall lines.

C. Slide veneer units into place and press firmly to achieve full contact of the mortar. Veneer units (brick) shall be lifted and removed occasionally to verify that full bedding has been achieved.

D. Provide ___ in. (____mm) wide mortar joints.

3.02.D Specify the desired joint width (usually $^3/_8$ in. $\pm$ $^1/_8$ in. (10 mm $\pm$ 3 mm)).

3.03 POINTING

A. Allow mortar to set fully prior to starting the filling of the mortar joints. Take special care so as not to move the installed brick units.

3.03 Mortar joints must be filled fully to prevent moisture penetration. Tooled concave joints provide the best weather resistance and are therefore recommended. Specify if another joint type is desired.

B. Point all joints with the approved mortar.

C. Tool all mortar joints to a slight concave unless otherwise called for on the Drawings.

3.04 CLEANING

A. Promptly clean the brick veneer units as the work progresses to minimize the need for final cleaning. Remove all cement and mortar stains from the face of the brick veneer units.

B. At the conclusion of masonry work, the masonry contractor shall remove scaffolding and equipment used in the work along with all debris, refuse and surplus masonry materials from the premises.

3.04 Care should be taken to keep the work clean during construction. This will significantly reduce the amount of cleaning required at the end of construction.

The Specifier should state the preferred method of cleaning the veneer surface (such as sandblasting, acid washing, etc.). Note that specific instructions on the application of these methods is required as sandblasting can damage some surfaces and acid washing can cause green staining to occur on some bricks.

3.05 SEALING

Upon completion of the cleaning operations, and their approval by the Architect, apply _______________ water repellent coating manufactured by _______________ in strict accordance with Section 07____and the Manufacturer's installation instructions.

3.05 Sealing may be specified here or in a separate section.

3.06 SCHEDULE

Provide a schedule when stone surface finishes, colors or textures vary for different locations.

3.06 Schedule may include special instructions; i.e. variable width head and bed mortar joints or other particular features.

5.2 GUIDE SPECIFICATION FOR ADHERED VENEER USING LATEX PORTLAND CEMENT MORTAR

The following Section provides a generic Adhered Veneer Using Latex Portland Cement Mortar Specification. The user must adjust the provisions for job specific applications.

1.0 CONTRACTORS OPTION

INSTALLATION OF ADHERED VENEER IN LATEX PORTLAND CEMENT MORTAR

2.0 ADHERED VENEER INSTALLATION USING LATEX PORTLAND CEMENT MORTAR

NOTE: Latex portland cement mortars can be used as a $^{1}/_{16}$ in. (2 mm) thick bond coat to bond veneer to a portland cement mortar bed that is still workable, as is customary when a neat cement bond coat is used. They can also be used on a cured (pre-floated) portland cement mortar bed or cementitious backer units. Under normal conditions, a minimum of 20 hours cure at 70°F (21°C) is adequate but longer mortar bed cures up to 10 days are desirable.

2.1 APPLYING LATEX-PORTLAND CEMENT MORTAR

Clean surface thoroughly. Dampen if very dry, but do not saturate.

Apply mortar with flat side of trowel over an area no greater than can be covered while mortar remains workable. Within ten minutes before applying veneer and using a notched trowel of type recommended by mortar manufacturer, comb mortar to obtain even setting bed without scraping backing material. Cover surface uniformly with no bare spots, with sufficient mortar to insure a minimum mortar thickness of $^{3}/_{32}$ in. (2 mm) between veneer and backing after veneer has been tapped into place.

2.2 SETTING ADHERED VENEER

Refer to mortar and latex manufacturers' directions.

Press veneer into freshly combed mortar, insuring mortar contact with veneer while maintaining accurate joint alignment and spacing. Keep a minimum of $^{2}/_{3}$ of joint depth open for grouting.

Thoroughly tap all veneer into place to obtain maximum contact of bonding mortar on the back of veneer and back mounting material so that there is an 80% minimum contact area.

Obtaining 100 percent contact with veneer may require troweling a layer of mortar on the back of each piece of veneer prior to firmly pressing it into the combed mortar.

3.0 INSTALLATION OF INTERIOR AND EXTERIOR BRICK PAVING

3.1 GENERAL

Installation of brick paving shall be over a cement mortar bed, properly prepared concrete slabs or properly prepared cementitious backer units.

3.2 APPLYING LATEX-PORTLAND CEMENT MORTAR

In addition to the provisions of Section 2.1, insert filler in expansion and control joints.

3.3 SETTING BRICK PAVING

Press brick into freshly combed mortar, insuring mortar contact with paving while maintaining accurate joint alignment and spacing. Keep a minimum of $^{2}/_{3}$ of joint depth open for grouting.

Thoroughly tap all brick into place to obtain maximum contact of bonding mortar on the back of brick and back mounting material. Average uniform contact area shall be not less than 80 percent. The 80% coverage shall be sufficiently distributed to give full support to the brick with particular attention to this support under all corners of the brick.

Obtain 100 percent contact with brick by troweling a layer of mortar on the back of each brick prior to placing on the combed mortar bed.

3.4 GROUTING OF BRICK PAVERS

Before grouting, all brick must be firmly set, and all spacers, strings, ropes, and pegs removed.

Consult mortar manufacturers recommendations prior to grouting.

4.0 BRICK SPECIFICATIONS FOR LATEX-PORTLAND CEMENT MORTAR

4.1 INTRODUCTION

Latex additives for use in thin-set portland cement mortars are designed to improve adhesion, reduce water absorption and provide greater bond strength and resistance to shock and impact. These additives allow some latitude in time, working conditions, and temperatures.

Basically, the latex additives are water emulsions which are added to portland cement mortars in place of water or replacing part of the water. The dry components-portland cement, dry additives, graded sand-must be preblended for use with the latex additive.

4.2 SCOPE

This specification describes the test methods and the minimum requirements for latex-portland cement mortar.

NOTE: A latex sold separately shall meet the requirement of this specification when mixed in accordance with manufacturers recommendations.

4.3 DEFINITIONS

Latex-Portland Cement Mortar: A modified portland cement thin-set mortar for the bonding of adhered brick to which a polymer has been incorporated in latex form or as a redispersable powder. When added in latex form it is added as a replacement for part or all of the gauging water.

Fast Setting Latex-Portland Cement Mortar: A latex-portland cement mortar specifically designed to obtain the minimum required shear bond strengths at a much faster rate than normal latex-portland cement mortar.

Non-Sagging Latex-Portland Cement Mortar: A latex-portland cement mortar specifically formulated to reduce slippage of adhered veneer on vertical installations.

5.0 SAMPLING AND TESTING PROCEDURES (As Required)

Sampling: Obtain not less than one gallon of the brand of latex to be tested and a sufficient quantity of prepared dry mortar mix.

5.1 STRENGTH OF ADHERED VENEER

5.1.A Shear Strength to Adhered Veneer

Preparation of Specimens: Prepare 1000 grams of mortar. Assemble 16 shear specimens from 32 pieces of veneer units with a $^1/_8$ in. (3 mm) mortar layer by bonding two pieces of veneer together for each specimen. The $^1/_8$ in. (3 mm) mortar bonding layer shall be established by placing $^1/_8$ in. (3 mm) thick T-bar spacers on the smallest veneer edge; be sure to rub all mortar off the under side of the spacers by sliding them back and forth along the edge. The top veneer unit shall be placed on top of the mortar layer and squeezed until a $^1/_8$ in. (3 mm) thickness or mortar is obtained. The veneer shall then be slid back and forth along the stainless steel spacers to assure proper thickness of mortar. The veneer shall be slid to assure $^1/_8$ in. (3 mm) offset. Care shall be taken in removing the stainless steel spacers so as not to cant or slide the veneer units in relation to each other or to alter the offset distance. Specimens shall not be individually handled for 24 hours and shall be treated with care until broken. Cure all specimens for seven days at 70-77°F (21-25°C), at 45-55 percent relative humidity. Use measured area of bond or assume an area of bond or 14.4 sq. in. (9300 mm^2) in calculating the strength value in psi. Load at a rate of 1340 pounds (608 kg) per minute.

7-Day Shear Strength: After 7 days dry cure, test 4 specimens in shear and record values.

Requirement: Average shear strength greater than 100 psi (689 kPa).

4-Week Shear Strength: Dry cure 4 specimens for 4 weeks. Test in shear and record values.

Requirement: Average shear strength greater than 150 psi (1034 kPa).

12-Week Shear Strength: Dry cure 4 specimens for 12 weeks. Test in shear and record values.

Requirement: Average shear strength greater than 150 psi (1034 kPa).

5.1.B Compressive Strength

Preparation of Specimens: Using mortar freshly prepared, fill three 2 in. (55 mm) cube molds, cover with polyethylene film for 24 hours. Cure an additional 48 hours in the cube molds. At the end of 72 hours, remove cubes from the molds and store at 70°F to 77°F (21°C-25°C) and 45 to 55 percent humidity for 25 additional days. On the 28th day after casting, obtain compressive strength on one cube, generally following the procedure of paragraph 10.6 ASTM Test Method C109. Determine compressive strength on the two remaining cubes by loading the cubes to 50 percent of the total strength determined on the first cube and releasing the load to zero. Do this twice, obtaining stress strain diagrams either automatically with a recording compression tester or by reading and recording the compression (deflection) of the specimen every 100 pounds (45.36 kg) of gross load where a deflectometer is utilized with a non-recording compression tester (See ASTM E111). Maintain loading rate specified in ASTM C190. Load the specimen a third time, again obtaining a stress diagram, by loading to failure.

Requirement: Compressive Strength not less than 3000 psi (20.7 MPa).

6.0 PACKAGE LABELING

Application: The container shall be clearly labeled. The necessary directions for application to be used and the general instructions required below shall appear on the container.

Instructions for storage: Instructions for proper storage of the latex shall be given, including any cautions against damage by freezing that are required.

Use: Instructions for proper use with kinds of veneer, recommended tools and procedures for application shall be given.

Shelf Life: Each container of latex shall be labeled with date of manufacture and the age after which the latex cannot be safely used.

5.3 GUIDE SPECIFICATION FOR ANCHORED BRICK VENEER

GUIDE SPECIFICATION

PART 1- GENERAL

1.01 GENERAL REQUIREMENTS:

Division 1 requirements, Drawing, General Conditions, Supplementary General Conditions, and Special Conditions apply to this section.

1.02 WORK INCLUDED:

A. This section includes labor, material, equipment, tools and appliances for the complete installation of anchored veneer in the areas shown on the Drawings.

B. Installing, pointing and cleaning of the veneer.

C. Attaching of anchorage to supports provided by others.

D. Placing of anchorages in the veneer as detailed.

E. Setting into the veneer all bolts, anchors, metal attachments and inserts as furnished and located by others.

F. Removal of surplus veneer material and waste after completion of the veneer work.

1.03 RELATED WORK SPECIFIED ELSEWHERE

A. Reinforcing Steel: Section 04____.

B. Metal Accessories: Section 04____.

C. Mortar: Section 04____.

D. Rough Carpentry: Section 06____.

E. Caulking, Sealant and Waterproofing: Section 07____.

NOTES TO SPECIFIER

1.01 Division 1, General Requirements, is that portion of the CSI Format for Building Specifications which contains all provisions pertaining to the job as a whole, including the "Special Conditions".

1.02 and **1.03** "WORK INCLUDED" and "RELATED WORK" are listed here to define these facets, though they are sometimes not made part of the Division Specifications.

If these paragraphs are included, the Specifier must check that there are no portions overlapping or not covered elsewhere.

1.04 QUALITY ASSURANCE

1.04 Specify any special submittals or tests required to assure the materials and construction procedure will be of the quality desired.

A. Submittals:

1. Samples: Submit ______ veneer units of each type, including special shapes required to show range of colors, textures, finishes and dimensions.

1.04.A.1 Specify the number of veneer units required for approval. Coordinate submittal requirements with standard procedures.

2. Certifications: Furnish manufacturer's certification that the veneer units provided meet or exceed the requirements of this Specification.

B. Sample Panels:

1.04.B Sample panels and mock-ups are a good way to assure that workmen know what is expected for them to provide. The sample panels may be part of the finished project.

1. Construct a sample panel of the work of this Section at the location on the side indicated by the Architect.

2. Provide on sample panel for each combination of veneer patterns, bond pattern, and mortar color.

3. Sample panels shall be ____ ft. (___m) x ____ ft. (___ m).

1.04.B.3 Specify the size of sample panel required along with any special requirements on its construction. A 4 ft x 4 ft (1.2 m x 1.2 m) sample panel is common unless special conditions require otherwise.

4. Show color range, texture range, mortar color, joint tooling, bond pattern, cleaning, quality of workmanship and other aspects of the work relating to this section.

5. Make necessary adjustments in the sample panels and secure the Architects approval. Construction may not proceed until the Architect approves sample panel.

6. Sample panels shall remain on the project site to serve as a datum for comparison with the remainder of the work of this Section for the purpose of acceptance or rejection.

7. If the sample panel is not included in the actual masonry work, it shall be demolished after the masonry work for the project has been accepted.

1.05 PRODUCT DELIVERY, STORAGE AND HANDLING

1.05 The intent of the Section is to keep materials clean, dry and in an acceptable condition during and after job delivery.

A. Store veneer units, cement, lime and other materials of this section off ground to prevent contamination with mud, water or other harmful materials.

B. Veneer units shall be handled and stored in such a manner so as to prevent chipping, breaking and cracking.

C. Cover and protect materials from inclement weather to maintain quality control and physical requirements.

PART 2 - PRODUCTS

2.01 VENEER UNITS

A. The anchored veneer shall be ___________, as produced by ___________.

2.01.A The Specifier must call out the type of veneer and desired supplier.

This will have been decided by preliminary samples and type of work previously determined by the Designer and the Supplier.

B. Reference Specification: ASTM C ______.

2.01.B See Chapter 2, Section 1 for applicable ASTM Guideline.

C. Dimensions: ____ in. (___ mm) wide by ____ in. (___ mm) long by ____ in. (___ mm) thick.

D. Provide all units with straight cut edges and square corners unless otherwise shown or specified.

E. Do not permit veneer units to vary more than $^1/_8$ in. (3.2 mm) in any dimension from those specified.

Provide units similar in texture and color to those approved by the Architect.

2.02 MORTAR

Mortar shall comply with the requirements of Section 04100

2.02 Mortar is often specified here. The editors choose to make a separate mortar specification for simplicity.

2.03 GROUT

A. Grout, if used between the veneer units and the backing, shall be 1 in. (25.4 mm) minimum thickness of cement grout.

B. Grout shall be fine grout.

1. Fine grout shall be 1 part portland cement and 3 parts sand.

2.04 OTHER MATERIALS

2.04 Specify any other materials pertinent to the particular job.

A. All ties and anchors shall be of non-corrodible material or corrosion protected material in compliance with local Building Code requirements.

B. "Paper-backed mesh" shall consist of waterproof paper or a membrane equivalent to asphalt saturated felt free from holes and breaks and weighing not less than 14 pounds per 100 sq. ft (0.68 kg/m^2). Felt shall be applied weatherboard fashion, lapped not less than ____ in. (___ mm) at horizontal joints and not less than ____ in. (___mm) at vertical joints. Over this apply wire mesh not less than 16 gauge (1.575 mm) at ____ in. (___ mm) o.c. with provisions for furring the mesh from the membrane, for embedment of wire.

2.04.B Specify type of paper-backed mesh along with lapping requirements and any additional requirements.

PART 3 - EXECUTION

3.01 WORKMANSHIP

A. Examine the areas and conditions under which work of this Section will be performed. Correct conditions that are detrimental to timely and proper completion of the work. Do not proceed until unsatisfactory conditions are corrected.

B. Veneer shall be laid in a pattern as shown on Drawing number ____.

C. Veneer shall be clean and free of dust and shall be laid with not less than 1 in. (25 mm) of air space, mortar or cement grout between the veneer and the backing wall. The space behind the veneer shall be solidly filled with mortar or cement grout or left void with spot bedding as required.

3.01.C The space between the veneer and backup may be solid grouted or left void as a cavity with provisions for weep holes.

D. All joints shall be filled solidly with mortar. Provide tooled concave mortar joints unless otherwise specified or noted on the Drawings. Tooling shall be done when the mortar is partially set but still sufficiently plastic to bond.

3.01.D Tooled concave joints provide the best type of joint to prevent water penetration. Specify if another joint type is desired.

E. No veneer unit shall exceed 10 in. (254 mm) in thickness at any point measured from back to face as laid. When veneer is placed in its bed joint, it shall remain at rest before the remaining mortar is applied around and behind.

3.01.E Depth of veneer, including mortar and grout, is recommended to be a minimum of 3 in. (76 mm) to a maximum of approximately 11 in. (279 mm).

There shall be no outside support other than the veneer beneath it and the anchor ties to the backing wall. In no case shall the overall distance from the backing wall to the exposed face exceed 11 in. (279 mm).

In some instances, thicker light weight veneer may be used. The prescribed limits have been selected merely because of customary use in local codes.

F. Veneer ties shall have adequate corrosion protection or at least $^5/_8$ in. (16 mm) of mortar coverage.

G. Wire mesh shall be attached to studs as shown on Drawings.

H. Anchors shall be installed to provide not less than the support shown on the details and in such a manner as to eliminate looseness or lack of fit of dovetails in slots, improper bends or kinks in wire, etc.

3.01.H Details of veneer anchors and other veneer requirements should be shown on the Drawings.

The designer should detail the type of anchorage of the veneer. These should show clearly the number, type, method of attachment, etc. The basic requirements for ties is that they be permanently adequate, of a material proven to be corrosion resistant. Ordinary reinforcing bars or ordinary steel wire in joints, when protected by the cement mortar and grout have generally proven to be satisfactory.

Anchors that will be exposed to a corrosive atmosphere must be galvanized, coated or of stainless steel, or other equivalent protection.

3.02 CLEANING

A. Mortar stains shall be removed with clear water as work progresses. Upon completion, all exposed surfaces shall be cleaned removing all stains with fiber brushes and then washing with clean water.

3.02.A Specify the desired cleaning method. Note that a careful selection is necessary as acid washing can discolor some brick and stone and sandblasting can harm the surface of many materials.

B. In the event ordinary cleaning is not adequate, the Contractor shall use special cleaning methods such as sandblasting, acid washing, chipping, etc., as approved by the Architect.

C. At the conclusion of masonry work, the masonry contractor shall remove scaffolding and equipment used in the work along with all debris, refuse and surplus masonry materials from the premises.

3.03 SEALING:

Upon completion of the cleaning operation, and their approval by the Architect, apply __________ water repellent coating manufactured by ______ in strict accordance with Section 07 ____ and the Manufacturer's installation instructions.

3.03 Sealing may be specified here or in a separate Section.

5.4 GUIDE SPECIFICATION FOR EXTERIOR STONE CLADDING

The following Section provides a generic Exterior Stone Cladding Specification. The user must adjust the provisions for job specific applications.

1.00 GENERAL

1.01 DESCRIPTION: Division 1 applies to this Section. Provide exterior facade stonework as indicated, specified and requested.

A. Work in this Section: Furnish and Install Exterior Facade Stonework in accordance with requirements of the contract documents. This work includes, but is not limited to, the following.

1. All exterior stone facing panels, special shapes, piecework and all other stone element for the building facade.

2. Quarrying, fabrication, drilling, fitting and cutting of stonework as required for the proper completion of the work.

3. Stonework set to units or steel frames, on or off the job site prior to installation including stone anchorage, water control gutters only in truss frames, and weep drains as specified hereinafter. *(Consideration for moisture control should be noted by specifier).*

4. Handset stonework on CMU, cast in place concrete, hot rolled or cold formed steel backup systems including stone anchorage and vent and weep drains as specified hereinafter. *(Backup waterproofing and flashing should be pre-installed by appropriate trade).*

5. Testing as specified herein, including but not limited to full scale testing of individual panels, physical property testing, anchorage testing, integrated wall system testing, and sealant compatibility testing.

6. Delivery of stone to job site or wall assembly facility including protection during transit.

7. Samples for visual quality control samples at job site. (*The quarry may elect to maintain samples for quality control*).

8. Accessories and hardware required for complete installation.

9. Shop drawings, engineering calculations, erection drawings, and any special tag or marking drawings required.

10. Leave stone clean after installation.

11. Make provisions for anchors, including slotted type, dowels, clamps, rods, hangers, clips, ties, bolts, or other fastening devices as required to securely attach the stone to the support material or to the structure. Where such items are to be built into or cast into the support material or the structure by other sections, coordinate same and provide locations drawings.

12. Surplus stock as required.

13. Visual mock-up at the job site.

B. Related Work Specified Elsewhere

1. Steel Frame Support System for Stone Veneer.

2. Cold Formed Metal Framing System for Stone Veneer.

3. Facade Sealant.

4. Facade Insulation and Firesafing.

5. Fixed Windows.

6. Mock-up.

7. Flashing and Waterproofing.

1.02 QUALIFICATIONS

A. Engage a fabricator with a minimum of 5 years successful experience in the fabrication of stone of similar sizes, shapes and finishes to the units required for this project, and which has ample production facilities to produce and furnish the units as required without delay to the work.

Engage an installer with a minimum of 5 years successful experience in the erection of stone of similar sizes, shapes and finishes to the units required for this project, and which has facilities to install the units as required without delay to the work.

B. Stone suppliers shall provide stone range samples at their facility for review by the design team. The stone suppliers shall provide three 24 in. x 24 in. (610 mm x 610 mm) samples and finish specified including historical testing data for each.

C. Quarrying shall be supervised and coordinated by the stone fabricator to assure that the as–quarried block orientations will yield finished material with characteristics as described herein. Only one source of each type of stone shall be used throughout the work.

D. The Contractor, by commencing the work of this section, assumes as a part of his warranty of the work, to assure that components and parts shown or required within the work of this section and comply with the contract documents. The Contractor shall further warrant that all components, specified or required to satisfactorily complete the installation are compatible with each other and with the conditions during installation and the expected use. He shall warrant the overall effective integration and correctness of individual parts and the whole of the system.

E. All pieces of stone shall be within the approved range samples. Stone rejected for non-conformance to approved range samples or approved shop drawings shall be replaced with material meeting the specified requirements.

F. Stone shall match the type, pattern, color, texture and finish of samples available for inspection in the office of the Architect.

G. Architect's inspection of the stone does not relieve the Contractor from his responsibility to provide all stonework in accordance with the approved contract documents, samples and shop drawings.

H. All examinations, selections and approvals shall be for the purpose of achieving a final appearance of stone with the greatest possible uniformity, and will be based upon the following criteria:

1. Color and value within approved, pre-selected color ranges and finishes.

2. Finish, within approved, pre-selected ranges of variation.

3. Uniformity of texture and graining within approved, pre-selected ranges of variation.

4. Conformance to approved shop drawings and details with specified dimensions and tolerances.

(Any other special requirements should be noted).

I. Quarried blocks shall be made available for inspection by the Architect at his request.

J. Production units shall be made available for inspection by the Architect at his request. To this end, the fabricator shall advise the Architect when production has begun and of the earliest possible opportunity to inspect a representative sampling of production work.

K. All examinations, selections and approvals shall be for the purpose of achieving a final appearance of stone with the greatest possible uniformity. Rejection will be based upon the following criteria:

1. Color and value not within approved typical full size, pre-selected color range and finish samples signed by Architect and located at the quarry and job site.

2. Graining, from finishing work, in excess of reasonable minimum as approved by the Architect, and not running generally parallel on all finished surfaces.

3. Non-conformance to approved shop drawings and details within specified dimensions and tolerances.

4. Any stone with a crack may be rejected and require replacement, as determined by the Architect.

1.03 PLANS AND SPECIFICATIONS

A. The character of these requirements is intended to provide a performance type specification for the design, fabrication and installation of the Exterior Facade Stonework. The Contractor is responsible for the engineering and design of all components and materials as well as the fabrication, installation and performance of the Exterior Facade Stonework.

B. Drawings are diagrammatic. The details shown are intended as a guide for the aesthetic and interfacing requirements of the stone and its support system. The requirements shown by the details are intended to establish basic dimensions of the module and the sight lines, jointing and profiles of members. The Contractor is responsible for the design and engineering of the system within these aesthetic parameters. The drawings are not to be construed as engineering design, nor necessarily adequate to meet the engineering design requirements.

C. It is recognized that the design details do not cover some conditions or modifications which may be required. It is, however, intended that conditions not detailed shall be developed through the Contractor's shop drawings to the same level of aesthetics and in compliance with performance criteria as indicated for detailed areas and as stipulated in these specifications.

D. All stone projections and coping to slope for drainage.

E. Provide water drip cuts at all horizontal returns and soffits.

1.04 CODES AND STANDARDS

The work, except as otherwise shown or specified, shall comply with the latest edition of the following. Stone and workmanship quality shall be in accordance with the best industry standards and practice as set forth by the following. When conflicting requirements arise, follow the more stringent requirement.

A. Standards of the American Society for Testing and Materials (ASTM Standards).

B. National Building Granite Quarries Association (NBGQA).

C. American National Standards Institute (ANSI).

D. "Local Building Code, current edition."

E. Marble Institute of America.

F. Building Stone Institute (BSI).

G. Indiana Limestone Institute (ILI).

1.05 SUBMITTALS

A. Submit shop and fabrication drawings stamped by a state licensed professional engineer for exterior facade stonework.

This engineer shall also perform and submit structural calculations to document all stone stress conditions. Coordinate drawings and drawing submittals with anchorage system. Show in detail the jointing, anchors and dowels, if any, dimensions, sizes and locations of cut-outs, relation to work of other trades, and all other pertinent data and information. Detail all stone, shapes, thickness, dimensions of slots or holes for anchors, finishes, miter and corner conditions. Coordinate stone sizes with window, curtain wall, and ground floor entrances and storefront details to ensure adequate sealant coverage at perimeters. Number each piece of stone to correspond to the markings as shown on the fabrication drawings. Mark the identification number on the back of each stone.

B. The Contractor shall prepare and submit performance mock-up drawings (if required) showing the scope, method of attachments, finishes, edge finishes, thickness, tolerances, and any other pertinent information. Mock-up drawings shall be accompanied by structural calculations for stone strength and anchorages shown in drawings. Both drawings and calculations shall be stamped by the same engineer. Engineer must be licensed in the state where the job is located. Drawings shall be incorporated into the exterior wall drawings. The drawings shall then be included in the coordinated mock-up drawings with all other trades as prepared by the Contractor.

C. Provide duplicate 24 in. x 24 in. (610 mm x 610 mm) quality control sample slabs of stone in each finish required, indicative of the entire range of materials and respective finishes. Prior to fabrication, provide a visual display of these samples at the quarry or stone fabricators plant for the Architect's review and approval. These samples will serve as the control for limiting the acceptable range of appearance. The samples will be signed and dated by the Architect and the General Contractor. One set of signed stones shall be shipped to the job site for use as quality control samples. The other set of samples shall be retained at the quarry or stone fabricators facility for their quality control. Also, provide samples of facade stonework necessary for test mock-up and other testing, as specified hereafter.

D. Submit manufacturer's data sheet or equivalent printed literature indicating product information for stone, stone anchorages, epoxies, setting accessories and other related materials. Data shall substantiate that materials comply with specified requirements.

E. Submit test reports per ASTM E575 for results of the physical properties testing, as specified hereinafter. Submit test results of anchorage and full size panel tests as specified hereinafter.

F. Submit calculations for review by Architect, certified by a professional engineer licensed in the state where the job is located, for structural adequacy of stones and anchorage system. Coordinate calculations submitted with shop drawing submittals. Calculations to be stamped by said engineer. Analyses shall include seismic criteria as required by the local building code.

G. Submit stone fabricators quality control procedures regarding rift, thickness, cracks, finishing and other concerns.

1. After slabbing stone, measure thickness at four corners of one slab per block.

2. After finishing, measure thickness at four corners of every 20th slab. Remove any slabs that do not meet tolerance criteria specified herein.

3. Examine each slab surface for evidence of hairline cracks. Remove any slabs containing hairline cracks.

4. Wet surface of every 20th slab and observe while surface dries to find evidence of hairline cracks. Remove any slabs containing hairline cracks.

5. After fabrication, measure all dimensions of cuts, and drilled holes on every 20th panel.

6. General contractor's representative shall visit and review fabricators operations.

H. Submit stone repair procedures in a detailed outline of materials and methods one month prior to commencement of job site erections of any panels.

I. Submit stone replacement procedure details. Provide two replacement stone panels on test mock-up as selected by Architect. Replacement anchorage to be subjected to the same performance requirements as for original stone anchors.

J. The successful stone supplier shall provide, if necessary, a full size visual mock-up at the job site showing the full range of the approved stone(s). The scope of the mock-up shall be as indicated on the architectural elevations.

1.06 PERFORMANCE CRITERIA

A. Granite

1. Granite shall be tested in accordance with ASTM C99. Stone shall be tested for flexural strength per ASTM C880 modified.

2. Nominal granite thickness shall be a minimum 3 cm after finishing.

3. Theoretical structural calculations of granite stress shall be based on the lowest average of the four group averages of ASTM C99 or C880 modified tests. The working stress values shall have a minimum safety factor of 3.0 when compared to this average and are to be based on minimum section thickness net of tolerance after fabrication.

4. Adequate number and size of anchors for granite retention shall be provided to satisfy the load requirements and design criteria set out in this specification. In any case, anchors shall not be less in size and number than required by code. The ratio of the granite anchor ultimate strength to actual load at design pressure shall be at least 4.0. A $^{1}/_{3}$ increase in allowable stresses for wind is not allowed. Design for seismic loading shall be as required by local building code.

B. Limestone

1. Limestone shall be tested in accordance with ASTM C99. Limestone shall be tested for flexural strength per ASTM C880 modified.

2. Theoretical structural calculations of limestone stresses shall be based on the lowest average of the four group averages of ASTM C99 or C880 modified tests minus one standard deviation. The working stress values for limestone shall have a minimum safety factor of 6 when compared to this modified average. Calculations shall be based on the

minimum thickness tolerance section after finishing.

3. Adequate number and size of anchors for limestone retention shall be provided to satisfy the load requirements and design criteria set out in this specification.

The ratio of the stone anchor ultimate strength to actual load at design pressure shall be at least 8. A $^1/_3$ increase in allowable stresses for wind is not allowed. Design for seismic loading shall be as required by the local building code.

4. Nominal stone thickness for limestone shall be a minimum 2 in. (51 mm).

C. Methods of fabrication, anchorage, and assembly shall be in compliance with the project documents, however, they shall be at the discretion of the Contractor, with Architect's review, provided that the visible architectural effect is not changed.

D. Coordinate all stone anchorages with the design of the stone support systems.

1.07 TESTING

A. Testing of specified stones shall be performed by the Owner's independent testing laboratory located in the United States prior to all other testing programs. Testing to be completed prior to mock-up testing or fabrication of stone.

B. Identify finish, surface, rift and other necessary data for proper testing per the test procedures specified. Provide stone samples for each type and finish of stone as required by the approved independent testing laboratory for the following tests. Each different finish and stone type must be tested.

C. Prior to mock-up testing, prepare and test a minimum of six 12 in. x 12 in. (305 mm x 305 mm) stone specimens for each type of anchor design proposed in each direction of anticipated load, i.e. tension, shear, etc. Test set-up and load apparatus shall be submitted to the Architect or his Consultant for review prior to testing. Materials used and fabrication methods shall match those to be used for project production. The reports shall identify exact material specifications that were tested and dimensions for all components involved in testing. Load shall be applied at 10 psf increments, to ultimate failure. Deflections shall be recorded and reported at each load increment. Observation of initial yield shall be reported. Mode of ultimate failure shall be reported as well as load value at failure. Consistent failure mode shall be required for each type of specimen and load category. Deflection values in excess of 0.032 in. (0.8 mm) at 150% design load are not allowed. All granite specimens must exceed 4.0 times the design load without failure. All limestone test specimens must exceed 8.0 times the design load without failure.

Any failures which are less than the specified criteria shall require redesign at no cost to the owner.

D. For anchorage systems utilizing stainless steel pins set into stone and secured with epoxy materials the following criteria and testing procedures shall be maintained:

1. Stainless steel pins to be $^1/_4$ in. or $^5/_{16}$ in. (8 mm) dia. threaded full length with the exception of point at which bend occurs. Shop installation only shall be provided.

2. Pins shall be set into stone holes not exceeding $^1/_{16}$ in. (1.6 mm) greater diameter than size of pin used. Such holes must be drilled with diamond drill bits. If design is based upon a bent bolt type anchor, bolt shall be at $22^1/_2$ degrees from horizontal. A minimum clearance from bottom of hole to outside face of stone of $^1/_4$ in. (6 mm) must be maintained.

3. Prior to placement in hole, the hole shall be dry and completely cleaned.

Hole shall then be filled with a high modulus epoxy. Such epoxy shall be identified by name and manufacturer. Manufacturer's literature shall be supplied.

Contractor shall warrant that the epoxy formulation is satisfactory to provide adhesion to granite. The use of polyester based epoxies is prohibited except as filler materials for chip repair.

4. Sufficient time for epoxy to thoroughly set must be allowed prior to securing pins to connected elements. Bent bolts and nut shall be torqued to 4 ft lb. (5.42 N•meter) minimum, if this design is used.

5. Pins shall be tested in shear as follows: (Load vs. deflection readings shall be charted for each test specimen).

<table>
<tr><th>PHYSICAL PROPERTY</th><th>TEST METHOD</th><th>NO. OF TESTS</th></tr>
<tr><td>1. Compressive Strength</td><td>ASTM C170</td><td>3</td></tr>
<tr><td>2. Modulus of Rupture
Tested Dry Parallel
Tested Wet Parallel
Tested Dry Perpendicular
Tested Wet Perpendicular</td><td>ASTM C99</td><td>5
5
5
5</td></tr>
<tr><td>3. Density</td><td>ASTM C97</td><td>3</td></tr>
<tr><td>4. Absorption
(% of weight)</td><td>ASTM C97</td><td>3</td></tr>
<tr><td>5. Flexural Strength
Tested Dry Parallel
Tested Wet Parallel
Tested Dry Perpendicular
Tested Wet Perpendicular</td><td>ASTM C880
Modified*</td><td>5
5
5
5</td></tr>
<tr><td colspan="3">*Modify specimen size to be 9" wide x 38" long x job thickness with finished face loaded in flexural tension. Finish to be same finish as that used on the project. Each different finish must be tested.</td></tr>
<tr><td>6. Volumetric and
Chemical Stability</td><td>ASTM C295
Modified**</td><td>10</td></tr>
<tr><td colspan="3">a. **Select 10 pieces of used C880 or C99 specimens which display the lowest strength values as tested above for thin section examination of susceptibility to freeze-thaw damage, inherent crack exacerbation mechanisms, soluble compounds, and other items which affect durability or strength.

b. Conduct the petrographic examination on thin sections using petrographic and scanning electron microscopes and other instruments as needed. Use samples oriented parallel and perpendicular to the plane of stone panels and to the rift, noting orientations in the report. The source and method of selecting samples must be reported.

c. Provide estimated modal compositions of the stone. Neither quantitative chemical analyses of whole rock nor individual minerals or point counts are required. The presence of any mineral or feature that may adversely affect the durability or aesthetics of the stone must be reported. Provide data on texture, including the following.

i. Nature, density, and orientation of cracks and microcracks.
ii. Secondary mineralization.
iii. Nature and distribution of porosity.
iv. Nature of mineral intergrowths.
v. Extent and nature of alteration/weathering of minerals.
vi. Systematic fabric elements.
vii. Grain size and shape of mineral species.</td></tr>
<tr><td>7. Durability Testing</td><td>ASTM C67
Section 8
Modified***</td><td>10 Parallel
10 Perpendicular</td></tr>
<tr><td colspan="3">***Modify specimen size to be 9" (229 mm) wide x 38" (965 mm) long x job thickness; add testing of 5 parallel and 5 perpendicular specimens per ASTM C880 (modified) after completion of 50 cycles and after weight loss measurements. Compare results to initial flexural strength tests. If post freeze-thaw strength average is less than 80% of initial strength average then Architect may require additional thickness or redesign of cladding at no cost to the Owner. Continue testing for an additional 50 cycles and perform C880 modified tests and weight loss measurements. Apply same design criteria comparison as above for evaluation of results. (This test is not required in areas of mild climate, such as the Southwest United States.)</td></tr>
</table>

a. Apply basic load of 50# (222N), hold for ten seconds and release.

b. Re-apply load to 50# (222N), 75# (334N), 100# (445N), 125# (556N), 150# (667N) and 175# (778N) in increments pausing ten seconds between each increment.

c. Continue applying 25# (111N) increments and hold each ten seconds until failure.

6. Pins shall be tested in tension as follows: (Record deflections)

a. Apply tensile load starting at 50# (222N) and increase loads in 25# (111N) increments to failure, pausing ten seconds between each increment.

7. Samples tested shall be performed on project finished samples 12 in. x 12 in. (305 mm x 305 mm) x thickness of stone (minimum 30 mm after finishing for granite and 2 in. (51 mm) for limestone). Five samples shall be tested, dry, cut with rift that will be used on the job.

8. In addition to deflections, document mode of failure, and load at failure for sample condition.

9. Five (5) additional samples shall be tested without epoxy. A non-adhesive filler should be used in lieu of epoxy. A safety factor of greater than 1.0 is necessary for this condition for the anchor to be acceptable.

E. Submit stone samples, finished per the job conditions to the sealant manufacturer for testing of adhesion, compatibility and stain testing. Refer to Facade Sealant for further requirements.

F. Provide stone for the laboratory mock-up test specimen. Type, quantities and sizes to be same stone as stone for project. Stone must come from quarry location to be used for job stone. Scope to be suggested on the Architectural drawings but the responsibility of Contractor. Failure of stone or stone anchorage shall require replacement and/or redesign at no cost to General Contractor or Owner.

1.08 DELIVERY, STORAGE AND HANDLING

A. Contractor shall deliver stone to job site using protection and handling equipment which will ensure that there will be no breakage, chipping, soiling or other damage. Store above grade on wood dunnage or other suitable surfaces using polyethylene film to separate stone from wood or other supporting or protecting members. Protect from weather, soiling and damage of every kind. Crate stone to prevent accumulation of moisture between stones.

B. Any costs arising from breakage of stone or discoloration of stone during shipment, handling and installation shall be borne by the Contractor.

C. Furnish other materials, except bulk materials, in manufacturer's unopened containers with name, brand, type, grade and color fully indicated thereon. Store bulk materials as required to avoid any deleterious effects of weather, soiling or contamination.

D. All delivered items whether F.O.B. jobsite or fabrication facility and/or job site, for unloading by others, or whether fabricated and installed by the Contractor, shall be properly crated. Crates shall be marked with installation location, fabrication/piece numbers shop drawings reference, etc.... as applicable.

1.09 WARRANTY

A. Before final acceptance, the General Contractor shall guarantee to the Owner in writing that all parts of the work will meet the specified overall performance requirements of the specifications, and will be free from defects in materials and workmanship and remain watertight for a period of one (1) year from acceptance of the work by the Owner. The Contractor shall certify in writing that all work is in accordance with the contract documents and authorized alterations and/or additions thereto and that, should any defect develop during the guarantee period due to improper workmanship or materials under his jurisdiction, such defects will upon written request, be repaired or replaced at no additional cost to the Owner. If exploratory work is required to determine the cause of the defects, the cost of such work shall be borne by the Contractor only in case his work is found to be at fault.

B. Defective materials and workmanship are hereby defined to include structural failure of components resulting from exposure to normal load and forces, sealant failures and failure to fulfill other specified performance requirements.

C. The guarantee, the enforcement or lack of enforcement thereof, shall not deprive the Owner of other actions, rights or remedies available to him. Guarantee shall be in form approved by Architect. Sample warranty shall be submitted for review and approval.

2.00 PRODUCTS

2.01 MATERIALS

A. Stone

1. Provide stone of soundness (hardness and density) as required per the appropriate ASTM standard, and of texture, graining, color and tone, matching the samples in the Architect's field office and subject to the Architect's acceptance. Stone shall be free from defects which will impair strength, durability and appearance, and provided from a single quarry source to satisfy the total requirements of the project. Quarry and fabrication plant facilities shall be available for the Architect's inspection at any time. Physical properties shall comply with either ASTM C615 for granite or ASTM C568 for limestone.

2. Provide stones of type, color and finish as scheduled on the Architectural Drawings. Finish all exposed edges as scheduled.

3. Nominal granite thickness shall be a minimum of 30 mm. Nominal limestone thickness shall be 2.0 in. (51 mm). Increase thickness for aesthetics as required. Also, increase thickness as required relative to span provided by support system selected by Contractor.

4. The finish, range of texture uniformity, degree of graining, direction of grain drift and the range of color and value of all material shall match or otherwise conform to the properties as evidenced in quality control approval samples, sample quarry slabs, and full size mock-ups and pre-construction testing.

5. Stone quantities for each stone type to satisfy the total requirements of the project shall be provided from a single quarry as described herein.

6. Stone finish, as scheduled on the Architectural drawings shall be on all visible surfaces and exposed edge returns. Locations shall be as shown on drawings, or approved shop drawings. Degree of finish must meet with samples approved by the Architect.

7. Direction of grain drift shall run generally parallel on all finished surfaces. Erected work shall be oriented in direction as noted on approved shop drawings and approved by the Architect.

8. All stone shall be free from holes, seams, shakes, clay pockets, spalls, stains, starts, and other defects which would impair the strength, durability and appearance of the work, as determined by the Architect.

9. The following minimum criteria for quality control of stone cracks should be maintained:

   a. Cracks greater than 0.002 in. (0.05 mm) in width and 0.125 in. (3.2 mm) in depth are not acceptable.

   b. Cracks less than 0.002 in. (0.05 mm) in width can be no longer than 4.0 in. (102 mm).

10. All quarry cracks, seams, fissures, veins and other natural cracks are to be prequalified by the Architect as healed, partially healed or open. Partially healed and open cracks are not acceptable and should not be supplied to the project.

11. All stone shall be selected for background color, and veining. Marking and matching shall run in even shades and shall be set accordingly. Inherent variations characteristic of the stone and the quarry from which the stone is obtained shall have been brought to the attention of the Architect at time when the stone samples were submitted to the Architect. These conditions shall be subject to Architects approval.

B. Stone Types:

"List Stone Types"

C. Stone Accessories

1. Anchor bolts, pins, dowels, nuts and washers shall be AISI type 302/304 stainless steel if in contact with concrete or stone.

2. For stone on precast, rubber or neoprene grommets, tubes shall be of the diameter required to install around dowel pins that anchor into concrete. Wall thickness of tube to accommodate the differential volumetric expansion of the stone and concrete.

3. Epoxy Resins: Minimal field application of epoxy is allowed on this project.

a. Epoxy resins shall be selected from the following manufacturers:

i. Sika
ii. Adhesives Technologies
iii. Bonstone

b. The epoxy shall have a minimum heat deflection temperature of 145 degrees F. (63 degrees C.) per ASTM D648.

c. Epoxy used structurally shall only be used in conjunction with 300 series stainless steel mechanical retention mechanisms. Joints shall maintain structural stability under design loads in the event of total adhesion failure.

d. The epoxy shall not be considered as the primary retaining mechanism. The epoxy may be used to enhance the structural capacity of assemblies.

e. Epoxy manufacturer shall state in writing that the selected epoxy is acceptable for use with the stone and the application for which it is being used.

f. Polyester based epoxies are prohibited for structural applications. They may be used as fillers for chip repairs only.

4. Shims shall be stainless steel or aluminum or approved equal. Shim size shall distribute the loads so point loading does not affect stones performance.

5. Where permanent setting pads are required, use 90 durometer neoprene, lead or Korolath. Use of wood wedges is acceptable for temporary shimming only; they must be removed after installation is set.

6. Stone kerf details, if used, shall provide a minimum clearance of $^1/_{16}$ in. (1.6 mm) at all metal surfaces within kerf cut. Stone shall not bear directly against or on metal. Dead load of stone shall be isolated from metal supports by elastomeric materials. All kerfs shall be sealed with an elastomeric sealant prior to setting of kerf clip into stone kerf. Kerf cut shall be completely filled with sealant.

7. Anchorage systems which connect but which are not in direct contact with handset stonework shall be continuous galvanized or clip steel. They shall be anchored to hot rolled or cold formed steel back-up systems, CMU, or cast in place concrete with field installed anchors as required and approved at each condition. Continuous or clip deadload and wind load stone anchors shall be formed 316 stainless steel or extruded 6061-T6 aluminum members as required and approved and shall provide a minimum clearance of $^1/_{16}$ in. (1.6 mm) at all surfaces within kerf. All kerfs shall be sealed with an elastomeric sealant prior to setting of stone support members into stone kerf. Kerf cut shall be completely filled with sealant.

8. Provide vent and weep member at grade level of each major vertical stone joint. Vent and weep prefab assembly shall be $^3/_8$ in. x 1 in. (9.5 mm x 25 mm) high and of 316 stainless steel, and shall be set flush to exterior face of stonework.

2.02 FABRICATION

A. Quarrying shall be supervised and coordinated by the stone fabricator to insure that the as quarried block orientation will yield finished material with characteristics as described herein.

B. Accurately cut, dress, drill, fit and finish stone work to shapes and dimensions shown on drawings and/or final shop and setting drawings. Make exposed plane surfaces true in line. Work shall be in compliance with recommendations of the National Building Granite Quarries Association, Inc., the Marble Institute of America and the Indiana Limestone Institute.

C. Cut all joints and edges square and at right angles to face, and with backs parallel for face except at radial stone make joints radial. Make arises straight, sharp, true, and continuous at joints. No kerf cuts nor rebated cuts at exposed edges shall be permitted.

D. Cut, kerf, rabbet, back check and drill stones in shop as required for supports and anchors or for lifting as required by stone support system.

E. Allow for expansion and contraction within the limits of the joint material when cutting for anchorage devices. Back check stone as required for tolerances, and to accommodate imposed movements or stone anchorages.

F. All stone work shall be executed by mechanics skilled in the trade. Cut stone units with bed, unless otherwise approved by Architect.

G. Stone shall be accurately cut to sizes, shapes, profiles and dimensions. There shall be no deviation from jointing. Cut with rift in same direction throughout.

H. Stone kerf details, if used, shall provide a minimum clearance of $^1/_{16}$ in. (1.6 mm) at all metal surfaces within the kerf cut. Stone shall not bear directly against or on metal. Dead load of stone shall be isolated from metal supports by elastomeric materials. Stone support clips or elements shall be extruded aluminum with strength equal to or greater than 6063-T6. All kerfs in stone shall be completely filled with an elastomeric sealant prior to setting of kerf clip or kerf extrusion. Sealant to be compatible with perimeter sealant. Sealant to be tested in accordance with testing herein before specified.

I. Patching or repair of stone shall be performed only in accordance with the procedures approved by the Architect. No repair shall be done until procedures are approved. Chips or broken stones where breakages occur in non-exposed areas shall be considered repairable only if stone strength is not compromised and only if the area of repair is less than thirty–six (36) sq. in. (0.02 m^2) and the length of one dimension does not exceed 9 in. (0.23 m).

J. Make faces of stone units in the same plane flush at joints. All finished surfaces shall be true in line and face.

K. Saw-cut back surfaces which will be concealed in the finished work. Maintain mill finish on face of panels that form sealant joint from overlapping stone panel. Backs of stone units shall be sawn to true planes, parallel to face plane. Sawn surfaces and edges shall be cleaned of all rust stains and iron particles.

L. Dress joints (bed and vertical) straight and at 90 degree angle to face. Except at radial stone cut radial joints.

M. Provide stone panel of 30 mm after finishing for granite and 2.0 in. (51 mm) for limestone nominal thickness or thicker as required by test results. Provide greater stone thickness than shown where thickness shown is insufficient for sizes per test results or where extent of cutouts shown decreases effective strength of remaining material. The use of liners properly sized and secured to stone facing will be permitted only where shown on drawings and/or final shop drawings and shall be mechanically secured in addition to adhesive connection.

N. Provide reveals, reglets, openings, and other similar features or spaces as required for contiguous work. Coordinate with drawings and final shop drawings showing contiguous work.

O. Stone to receive anchors, dowels, and other necessary items. Lewis holes and holes for other back anchoring devices shall not be closer than 2 in. (51 mm) of edges.

2.03 TOLERANCES

A. Tolerances are not cumulative. The following tolerances shall be used for stone fabrication. Coordinate all tolerances with the stone support system.

1. Thickness: plus or minus $^1/_{16}$ in. (1.6 mm).

2. Hole diameter: plus or minus $^1/_{32}$ in. (0.8 mm).

3. Step in Face Dimension: plus or minus $^1/_{16}$ in. (1.6 mm).

4. Angle of Kerf Cut or Drilled Hole: plus or minus 2 degrees.

5. Depth or Width of Kerf Cut: plus or minus $^1/_{16}$ in. (1.6 mm).

6. Deviation from Squareness: plus or minus $^3/_{32}$ in. (2.4 mm).

7. Joint Dimensions between Stone: plus or minus $^3/_{32}$ in. (2.4 mm) for 1 in. (25.4 mm) joints and $\pm$ $^1/_{16}$ in. (1.6 mm) at $^3/_8$ in. (9.5 mm) joints.

2.04 FINISHES FOR STONE

A. No hiding of defects will be permitted. Units having flaws or imperfections are not acceptable and will be rejected. Replace with acceptable units without additional compensation. Patching of stone shall only be done in accordance with the specified criteria herein and with approval of the Architect (See 3.03 for repair of stone).

B. Examine finished stone for microcracking. Report instances of excessive or potentially damaging cracks in crystalline structure.

C. All exposed-to-view surfaces shall be finished. Exposed-to-view edges shall match face finish. Face finish shall be as noted on Drawings. If a surface finish is undefined, verify with Architect.

3.00 EXECUTION

3.01 INSPECTION/PREPARATION

A. Prior to installation, examine surfaces to receive stone and do not proceed until any defects detrimental to the finished work are corrected, including the moisture protection, structural supports, provisions for expansion, or any other condition which might affect the finished work in appearance, water tightness or integrity of the completed installation.

B. Verify all measurements and dimensions. Coordinate the installation of inserts for this work. Coordinate and schedule this work with the work of other trades. Give particular attention to the location and size of cut-outs required to accommodate other work or adjoining construction, in accordance with the reviewed shop drawings for such trade.

C. Review shop drawings of items or assemblies related to the support of anchorage for stone work, including requirements for clearances for proper installation.

D. Obtain written certification from the respective manufacturers of all proprietary products such as epoxies, sealants and anchorage devices that the materials are proper for the expected use and exposure of the stonework. No materials soluble in water after set shall be used. Submit data and documentation as confirmation to Architect for review.

E. Advise installers of other work about specific requirements relating to his placement of inserts or anchorages to be used by stone work. Furnish installers of other work with drawings or templates showing location of inserts for stone anchors and supports.

F. Coordinate all stone work and anchorages with the stone support system to produce work in compliance with the performance and aesthetic requirements set forth herein.

3.02 INSTALLATION

A. Install stone as shown and as detailed on Final Shop Drawings, or in compliance with recommendations of the National Building Granite Quarries Association, Inc., the Indiana Limestone Institute, and the Marble Institute of America.

B. Anchors shall be as detailed on approved shop drawings. All anchors which directly support stone shall be stainless steel. *(Alternate materials may be permitted).*

C. Erect all parts of the exterior stone wall plumb and true in proper alignment and relation to established lines and grades, with uniform joints and reveals and as shown on approved shop and/or erection drawings. Tolerances shall be as established in other wall sections, or as stated herein.

D. Do not install component parts which are defective in any way, including warped, bowed, abraded, chipped, cracked or broken. Remove and replace stone which has been damaged during installation or thereafter before the time of final acceptance.

E. Seal around all anchors, connections and penetrations with approved elastomeric sealant, which is compatible with the perimeter sealant. Prepare all joints for sealant application.

F. Seal around all pin and dowel connections that use epoxy. *(Field application of epoxy may be required).*

G. Laminated or composite assemblies of stone shall employ mechanical connections in combination with adhesive materials. The mechanical connections shall be the primary source of anchorage for all designs.

H. Erection Tolerances: The installed exterior stonework system components shall conform to the following erection tolerances.

1. Deviation for Vertical Members: $^1/_8$ in.(3.2 mm) maximum in one story height and $^1/_4$ in. (6.4 mm) maximum in a 30 ft (9.1 m) run and $^3/_8$ in. (9.5 mm) for height of building.

2. Deviation for Horizontal Members: $^1/_8$ in. (3.2 mm) maximum in a 30 ft (9.1 m) run and $^1/_4$ in. (6.4. mm) total for building.

3. Offset from true alignment between two abutting members: $^1/_{16}$ in. (1.6 mm).

4. Stonework joints, vertically and horizontally, below Level 2 shall be within $\pm$ $^1/_{16}$ in. (1.6 mm) of nominal joint size shown on documents and approved shop drawings.

I. Execute exterior stone work by skilled mechanics, and employ skilled stone fitters at the site to do necessary field cutting as exterior stone is set.

3.03 REPAIR OF STONE

A. It is the intent to have all exposed surfaces and edges of stone units free from cracks, broken corners, chipped edges, scratches or other defects affecting appearance. Repairs shall be made in accordance with limits established herein and/or with the Architect's approval.

B. Chips at the corner or edge of granite may be patched, providing the structural integrity of the stone is not affected and providing the patch matches the color and finish of the natural stone so that the patch does not detract from the appearance. In such cases, the Architect shall approve of such conditions.

C. Repair of limestone is permitted as described herein. Some chipping and cracking is expected; repair of small chips is not required if it does not detract from the overall appearance of the work or impair the effectiveness of weather seals as determined by the Architect. Where damage does occur, and it is judged to be repairable, industry practice permits repair of the affected stone. In such cases, the Architect shall approve such conditions.

3.04 SEALANTS

A. Sealants, backer rods, and related components shall comply with requirements of Facade Sealant.

3.05 CLEANING AND PROTECTION

A. Contractor to clean stonework after completion of work, using clean water and stiff-bristle brushes. Do not use wire brushes, acid type cleaning agents or other cleaning compounds with caustic or harsh fillers. Remove excess sealant by moderate use of mineral spirits or other solvent acceptable to sealant manufacturer and stone supplier, or other approved method.

B. Remove and replace exterior stone units which are broken, chipped, stained or otherwise damaged. Where directed, remove and replace units which do not match approved range samples.

3.06 ACCEPTANCE

A. Within the limits of the approved range samples, each and every piece of stone shall be subject to the Architect's approval, and any piece or pieces may be rejected after having been set or erected. Rejected stones shall be carefully removed and replaced with a new suitable stone without delay and without cost to the Owner. Any piece or pieces damaged in the removal and resetting of defective or rejected pieces shall also be removed, with new and acceptable pieces provided and set at no cost to the General Contractor or Owner.

B. Installed units which are chipped, cracked or otherwise damaged or which do not conform to the Specification requirements, shall be removed and replaced with acceptable material at no additional cost to the Owner.

5.5 GUIDE SPECIFICATION FOR CERAMIC VENEER

The following Section provides a generic Ceramic Veneer Specification. The user must adjust the provisions for job specific applications.

1.00 SCOPE

These specifications cover machine-extruded and mold-pressed ceramic veneer, made from clay and grog, and having a vitrified or glazed ceramic finish suitable for use on exposed exterior and interior walls and partitions. They do not cover sculpture.

2.00 PHYSICAL PROPERTIES

Ceramic veneer units shall conform to the physical requirements as prescribed in Table 1.

3.00 UNIT DESIGN

A. Adhesion Type Ceramic Veneer. The thickness of adhesion type ceramic veneer, including ribs, shall be not more than $1^1/_4$ in. (32 mm). Backs shall be either ribbed or scored.

B. Anchored Type Ceramic Veneer. The minimum net thickness of anchored type ceramic veneer, exclusive of ribs, shall be 1 in. (25 mm). Backs shall be either ribbed or scored.

C. Mold-Pressed Ceramic Veneer. The minimum thickness of the exposed faces of mold-pressed ceramic veneer shall be 1 in. (25 mm). Backs of special shapes shall be open and ribbed.

4.00 SIZE AND SHAPE

A. The face sizes and shapes of ceramic veneer shall be as specified (*see Standard Specifications for Furnishing and Installing Ceramic Veneer*).

B. Face Dimension Tolerances. The face dimensions (length and width) of machine-extruded ceramic veneer shall not vary more than $^1/_{16}$ in. (1.6 mm) over or under the dimensions specified on the setting drawings.

C. Warpage Tolerances. The exposed face of machine-extruded ceramic veneer shall not vary from a true plane by more than 0.005 in. (0.1 mm) per 1 in. (25 mm) of length.

5.00 PROPERTIES OF GLAZED AND VITRIFIED CERAMIC FINISHES

A. High-Fired Glazed and Vitrified Ceramic Finishes shall conform to the following requirements of imperviousness, chemical resistance and resistance to crazing.

1. *Imperviousness.* After the imperviousness test, no stain that can be seen from a distance of 5 ft. (1.5 m) shall remain on or beneath the surface, except a slight discoloration in the depressions of mottled finishes.

2. *Chemical Resistance.* The color of the glaze shall not change under the chemical resistance test.

3. *Resistance to Crazing.* The glaze shall not craze, spall or crack when subjected to one cycle or autoclaving in the crazing test.

4. *Adhesion.* Adhesion tests shall show no clean separation of finish and body, and the finish shall have portions of the body adhering to it over the complete exposed surface.

B. Low-Fired Ceramic Glazes shall show no etching or whitening of the surface under the acid and base test for low-fired glazes.

6.00 FINISH

Finished faces of ceramic veneer that will be exposed when in place shall be free from chips, crazes, blisters, crawling or other imperfections detracting from the appearance of the finished wall when viewed form a distance of 5 ft. (1.5 m).

7.00 INSPECTION

The purchaser or his authorized representative shall be afforded opportunity for sampling and inspecting units at the place of manufacture, prior to shipment. At least 10 days from the time of sampling shall be allowed for completion of tests. The expense of the inspection and testing shall be borne by the purchaser.

Table 1 – Physical Requirements

Designation	Minimum Compressive Strength Avg. of 5 Specimens		Minimum Compressive Strength Individual		Max. Absorption by 5-hr Boiling, Percent		Max. Saturation Coefficient*	
	psi	MPa	psi	MPa	Avg. of 5 Spec.	Individual	Avg. of 5 Spec.	Individual
Machine-Extruded and Mold-Pressed Ceramic Veneer	5000	34.5	3500	24.1	16	18	0.80	0.82

*The saturation coefficient is the ratio of absorption by 24–hr submersion in cold water to that after 5–hr submersion in boiling water.

5.6 GUIDE SPECIFICATION FOR FURNISHING AND INSTALLING CERAMIC VENEER

The following Section provides a generic Furnishing and Installing Ceramic Veneer Specification. The user must adjust the provisions for job specific applications.

1.00 SCOPE

These specifications cover furnishing and installing machine-extruded and mold-pressed ceramic veneer on exposed exterior and interior masonry walls and partitions. They do not cover sculpture.

2.00 SETTING DRAWINGS

The ceramic veneer manufacturer shall provide the required number of copies of scale shop drawings as detailed from the architect's drawings and approved by the contractor and architect. These shop drawings are to be used for setting and shall indicate all dimensions and sizes of joints, as well as all anchors, hangers, expansion and control or pressure-relieving joints, etc., which shall be furnished by others.

3.00 CERAMIC VENEER UNITS

A. Quality: Ceramic veneer shall conform to the requirements of *Standard Specifications for Ceramic Veneer* of the Architectural Terra Cotta Institute.

B. Texture and Color: The texture, ceramic finish and color of all exposed surfaces of ceramic veneer shall be approved by the architect, subject to normal ceramic variations. If polychrome or low-fired colors, such as gold, silver, orange or vermilion are required, the extent of their use shall be indicated on the architect's drawings or in the specifications.

C. Cutting and Fitting: All ceramic veneer shall be laid out at the factory and properly fitted and numbered to correspond with numbers on setting drawings. (If anchored, provisions shall be made in the ceramic veneer for the receiving of anchors, ties, etc.).

D. Packing and Delivery: All ceramic veneer shall be properly protected against damage in transit or at the building. The manufacturer shall deliver all the ceramic veneer in a clean and good condition, in quantity and in sequence as required to properly expedite the installation. Ceramic veneer shall be delivered f.o.b. truck at street curb (or at closest point accessible to trucks) at building site, or f.o.b. rail car's destination if so stated in contract.

E. Receiving and Handling: Setting contractor shall receive all ceramic veneer f.o.b. truck at closest accessible point to job site (or f.o.b. rail car's destination if so stated in contract) and be responsible thereafter for its good condition. He shall keep ceramic veneer units in original packing, standing on edge, one piece high, until ready for use. If delivered banded on pallets, he shall keep them intact until ready for use. Stacking of pallet units is not permissible. Material exposed to weather before setting shall be properly protected and covered to prevent wetting and resultant staining from wrappings or other materials. When stored on open ground, wood strips shall be placed under the ceramic veneer to keep it from contact with the ground.

F. Cutting: All necessary face cutting of ceramic veneer at the job site shall be done with a motor-driven abrasive saw, furnished and operated by the setting contractor.

4.00 MORTAR AND GROUT

Mortar shall be proportioned by volume for the type specified within the following limits:

A. Setting Mortar. 1 part portland cement, $^1/_2$ part high-calcium lime putty or Type S hydrated lime and 4 to $4^1/_2$ parts sand.

B. Grout for Anchored Type Ceramic Veneer

1. Mortar Grout. Mix setting mortar with sufficient additional water to cause the mixture to flow readily. This grout is used where the spacing between the ceramic veneer and the wall is between $^5/_8$ in. (16 mm) and $^3/_4$ in. (19 mm), which spacing is permissible for shapes not exceeding 1000 sq.in. (0.65 m^2) in surface area.

2. Pea Gravel Grout. 1 part portland cement, 3 parts sand and 2 parts graded pea gravel passing a $^3/_8$ in. (10 mm) sieve. This grout is used where the spacing behind the ceramic veneer is thicker than $^3/_4$ in. (19 mm).

C. Grout for Mold-Pressed Ceramic Veneer

1. Grout Filler. 1 part portland cement, 4 parts sand and 3 parts top gravel well graded.

D. Mortar Materials shall conform to the following:

Portland cement: ASTM C150, Type I.

Lime putty: Approved high-calcium slaked lime putty screened and aged at least 20 days, containing not more than 4 percent magnesium oxide.

Hydrated lime: ASTM C206, Type S.

Sand: Clean sharp sand, ASTM C144.

5.00 INSTALLATION

A. Supports and Anchors: Non-ferrous metal anchors shall be installed where shown on approved ceramic veneer setting drawings. Metal supporting or securing the ceramic veneer shall be thoroughly embedded in the masonry and shall be protected from corrosion by encasing in mortar or grout. In the case of concrete construction, loop dowel anchors or other anchors approved by the architect shall be used, together with horizontal grooves, 6 in. (152 mm) wide, cast in the concrete.

Note: Furnishing and setting of all anchors embedded in the concrete and the formation of horizontal grooves in the concrete shall be the responsibility of the general contractor under the concrete section of the specifications. Setting contractor shall furnish and install pencil rods and loose anchors described below and other anchors that may be required which are not provided by the general or concrete contractor under another paragraph of these specifications.

B. Wetting: At the beginning of setting each day, all walls to be faced shall be soaked with potable water, applied by a hose and spray nozzle, and shall be soaked again with water not sooner than 1 hr. before setting of ceramic veneer. Ceramic veneer shall be wetted in accordance with manufacturer's recommendations with potable water just prior to installation.

C. Setting:

1. Anchored Type Ceramic Veneer. Ceramic veneer shall be set true to line in setting mortar and shall be anchored in accordance with approved setting drawings. All spaces between anchored type ceramic and backing wall, $^3/_4$ in. (19 mm) in width, shall be filled solidly with anchored type, pea gravel grout. Spaces less than $^3/_4$ in. (19 mm) in width, shall be filled solidly with anchored type mortar grout.

2. Adhesion Type Ceramic Veneer. Just prior to application of mortar, a limited area of the wall and entire back of the piece of the ceramic veneer about to be set shall be given a brush coat of neat portland cement and water. Brush coats shall be of such consistency that they will spread evenly. Immediately thereafter, spread one half of the setting mortar coat on a limited area of the wall and the other half on the entire back of the piece of ceramic veneer, and tap it in place on the wall so as to completely fill all voids. The total thickness of the mortar coat shall average $^3/_4$ in. (19 mm), but sufficient mortar shall be used to create a slight excess which will be forced out at the joints and the edge of the piece when it is tapped into place.

3. Mold-Pressed Ceramic Veneer. Before setting in place, all units shall be turned bottom-side up and filled solidly with grout filler for mold-pressed ceramic veneer. After the fill has set sufficiently to permit handling, the units shall then be set and anchored as indicated in the setting drawings.

4. Soffits. When applied to soffits, each piece of ceramic veneer, in addition to the usual centers and wood wedges, shall be supported by suitable bent, vertical, wooden shores exerting a constant upward pressure until the mortar coat has set for several days.

D. Tooling and Pointing: If pointing is not specified, all mortar joints shall be tooled as the setting progresses to form a concave joint. When pointing is specified, all mortar joints shall be raked to a depth of $^1/_2$ in. (13 mm). Before pointing, joints shall be saturated with clean water, after which they shall be filled solidly with pointing mortar and tooled. All surplus mortar and stains shall be removed as the setting progresses and the surface shall be left clean.

E. Cleaning: Upon completion of the setting, all surfaces shall be cleaned with soap or washing powder detergent and rinsed thoroughly with clean water. Abrasives or metal tools shall not be used.

6.00 INSPECTION (CERAMIC VENEER MANUFACTURERS)

In accordance with project requirements.

7.00 SPECIAL REQUIREMENTS TO BE INCLUDED IN OTHER TRADE SECTIONS

In accordance with project requirements.

5.7 GUIDE SPECIFICATION FOR VENEER TILE INSTALLED ON EXTERIORS IN PORTLAND CEMENT MORTAR

The following Section provides a generic Veneer Tile Specification. The user must adjust the provisions for job specific applications.

1.00 SCOPE

Furnish all labor, materials, equipment, tools, transportation and services required to complete the installation of veneer tile work indicated on drawings and a specified herein.

(List scope of work).

2.00 WORK SPECIFIED IN OTHER SECTIONS

A. All surfaces to receive veneer tile over wood or steel stud construction shall be left with a scratch coat applied to a thickness of not less than 1/4 in. (6 mm) nor more than 1/2 in. (13 mm) from the structural face. Scratch coat shall be mixed in the proportions of one part portland cement to three parts clean sharp graded sand.

Wood or steel studs shall be covered with one of the following waterproofing materials: 15 lb. (6.8 kg) asphalt saturated building felt; ASTM D962 or polyethylene sheeting 0.004 in. (0.1 mm) thick, ASTM C156.

Wire reinforcing shall be furred out 1/4 in. (6 mm) from the studs and shall be 3.4 galvanized metal lath or 2 in. x 2 in. (51 mm x 51 mm)–16/16 or 1 1/2 in. x 2 in. (38 mm x 51 mm)–16/13, ASTM A82 and A185.

B. All concrete surfaces to receive veneer tile shall be heavily sandblasted to remove surface oil and slickened surface caused by forms of troweling.

C. After concrete surfaces are sandblasted they shall receive a dash coat mixed in proportions of one part portland cement to one and one half parts sharp fine graded sand. Add one part of Anti-Hydro to five parts of water and in turn add this into dry mixed sand and cement to bring to proper consistency. Roughened surfaces shall be cleaned and well saturated with water prior to application of dash coat. The dash coat shall provide an irregular coating over the concrete in thicknesses varying from a minimum of 1/16 in. (1.6 mm) to a maximum of 1/4 in. (6 mm). Maintain moisture by covering or repeated wetting, for a minimum of three (3) days, until cured to maximum hardness and bond. If a thicker application is needed use a scratch coat as listed in Paragraph 2.00 A.

D. Examination of Surface to Receive Veneer Tile. Representatives of the general contractor, plastering contractor and tile contractor firms shall test hardness of scratch coat or dash coat with hammer and chisel and agree to suitability before installation is started.

3.00 MATERIALS

A. Veneer Tile shall be of highest quality tiles of sizes and colors hereinafter specified, of size and patterns indicated on drawings. Veneer Tile shall be (state brand to be used) or approved equal. Tile shall be as perfect as it is possible to manufacture, and shall be certified satisfactory for intended use by the manufacturer. The colors and shades shall be reasonably uniform. The exposed face of the tile shall be a smooth even surface and uniform in texture without chips.

Tiles, as defined in ASTM C242 (Tile. — A ceramic surfacing unit) shall comply with ANSI A137.1.

Manufacturers or suppliers of tiles not covered by ANSI A137.1 must furnish proof of the adequacy of their product.

1. Size: All tiles shall be of uniform thickness and shall be (*give size*) size.

2. Colors: Shall be (*state colors*). All color used in the work shall match approved samples.

B. Portland Cement shall conform to ASTM C150, Type I.

C. Hydrated Lime shall be for masonry purposes and conform to ASTM C207, Type S, plus the added requirement limiting the unhydrated oxides to 8% maximum.

D. Sand shall comply with "Standard Test Method for Organic Impurities in Fine Aggregates for Concrete," ASTM C40 and "Specifications for Aggregate for Masonry Mortar," ASTM C144.

E. Sand – Extra Fine shall conform to Ceramic Tile Institute (CTI-70-6).

F. Water shall be free from any impurities that is injurious to the construction and shall be fit for human consumption.

G. Grout shall be one part of graded extra fine sand added to each two parts of portland cement for natural colored grout, – or grout shall be (*state color and give names of manufacturers for colored grout*).

H. Pure Coat shall be pure portland cement mixed with water into a trowelable wet slurry.

I. Mortar Setting Bed shall be mixed in the proportions of one part portland cement, one part hydrated lime, six parts of clean sharp sand.

J. Sealant for expansion joints shall be (*state color, type and manufacturer*).

4.00 WORKMANSHIP

Veneer tile shall be set and grouted by journeymen tile setters experienced in setting tile of this type under similar conditions.

A. At no time shall the mortar setting bed be over $^3/_4$ in. (19 mm) thick. Apply a plumb scratch coat where necessary to establish this. Plumb scratch shall be mixed in the proportions of one part portland cement to four parts clean, sharp sand and shall be let to stand a minimum of twenty-four hours before applying the mortar bed. Maintain moisture in plumb scratch by covering or repeated wetting, continue until setting bed is applied or for three full days if setting bed is applied or for three full days if setting bed is not applied prior to this.

B. Apply rod and float the mortar setting bed to a uniform plumb and level surface, allowing room for veneer to bring finished surface to required plane. Thickness of mortar shall be $^1/_4$ in. (6 mm) to $^3/_4$ in. (19 mm) as required, cut through the setting bed, back to the scratch coat, horizontally and vertically every 24 in. (610 mm).

C. Before positioning, apply to each tile a layer of neat cement paste $^1/_{32}$ in. (0.8 mm) to $^1/_{16}$ in. (1.6 mm) thick. Neat cement shall also be applied, as a skim coat, to the setting bed and then tile pressed and tapped in place to provide complete coverage between mortar bed and contact surface tiles without voids.

Alternate Thin–set Method: For this method delete–Paragraph 4.00 C – and replace as follows:

The exposed back of each tile shall be given a $^1/_{16}$ in. (1.6 mm) thick coat of the thin–set material. Tile shall then be placed in position on tested thin-set portland cement mortar, or rubber latex portland cement mortar, freshly combed onto the mortar setting bed with a notched trowel. All combing must be done in a horizontal direction. Care shall be taken to cover the mortar with the backbuttered unit before it has become glazed over from exposure to sun or wind. Units shall then be tapped firmly into place, true and even with the finished surface line or plane.

D. Where they occur and at least every 20 ft (6.1 m) in both directions in the structure, expansion joints or control joints shall be continued through the mortar bed, veneer tile, reinforcing wire and kept free of mortar and grout. These shall be filled with an approved caulking compound, as close as possible to the color of the grout.

E. Fill joints with grout. Compress and tool the grout in the joints so that it bonds with the setting bed and tile edges without voids. Joints shall be neat and flush with the surface of the tile.

F. Promptly clean tile as work progresses to minimize final cleaning. Do not leave mortar scum or grout to dry on tile faces.

5.00 CURING

A. Maintain dampness for three days. If necessary sponge down or fog spray.

6.00 CLEANING

A. Final cleaning: All cement and grout shall be removed from the face of the tile; then, when clean, polish dry.

B. Acid cleaning shall not be used on glazed tile. For unglazed tile acid cleaning shall not be done before 10 days after setting. Wet with water before cleaning with sulfamic acid solution. Protect all metal with grease. After acid cleaning flush thoroughly with clean water.

6 GLOSSARY OF TERMS

A

Absorption – The amount of water that a solid or hollow unit absorbs, when immersed in either cold or boiling water for a stated length of time; expressed as a percentage of the weight of the dry unit.

Abrasive Finish – A finish applied to fabricated stones that gives an even non-reflective surface.

Abrasive Hardness – A measure of the wearing qualities of stone for floors, stair treads, and other areas subjected to abrasion by foot traffic.

Adherence – The property of materials bonded to each other without mechanical anchorage.

Adhered – Veneer secured and supported through adhesion to an approved bonding material applied over an approved backing.

Admixture – Any material other than water, aggregate, portland cement, or lime that is used as an ingredient of mortar or grout to improve the quality and/or property of the material.

Agate – A variegated variety of quartz showing colored bands or other markings (clouded, moss-like, etc.).

Aggregate – Inert particles such as sand, gravel, and/or rock, which, when bound together with portland cement and water, will form a homogeneous system such as grout or concrete.

ASTM International – A voluntary organization that sets standards for testing and materials based on a consensus agreement.

Anchor – Metal rod, wire, or strap that secures building veneer stone or other masonry to structural framework, backup wall, or other elements, or holds stone units together.

FIGURE 6.1 **Strap anchor with drop dowel in sandstone anchored to concrete.**

Anchor Tie – Any type of mechanical fastener used to secure masonry to a support and stable system, such as a backup wall; usually used in tension application.

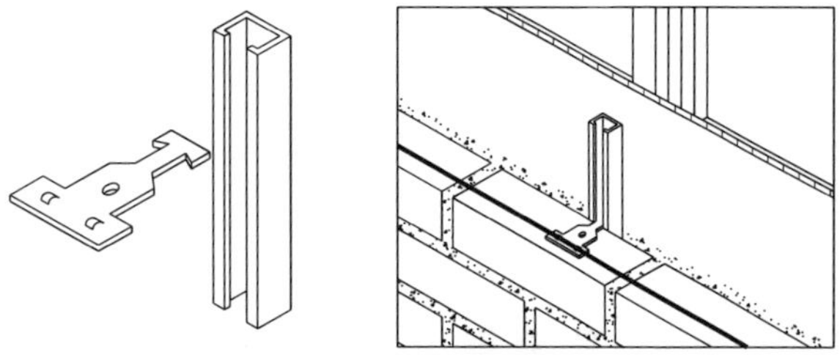

FIGURE 6.2 **Anchor veneer system.**

Architrave – The member of an entablature resting on the capitals of columns and supporting the frieze.

Argillite – A compact sedimentary rock composed primarily of clay and aluminum silicate minerals.

Arkose – A sandstone containing 10 percent or more caustic grains of feldspar. Also called ***Arkosic Sandstone, Feldspathic Sandstone***.

Arris – A natural or applied line on the stone from which all leveling and plumbing is measured.

Ashlar – Masonry having a face of square or rectangular stones, either smooth or textured.

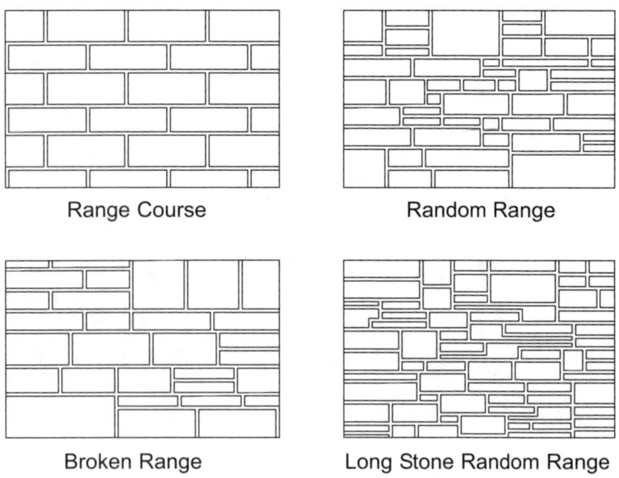

FIGURE 6.3 **Ashlar stone patterns.**

B

Backup – The part of a masonry wall behind the exterior facing.

Back–Buttering – Spreading mortar on stone units with thin-set material.

Baluster – A miniature pillar or column supporting a rail; used in balustrades.

Basalt – A dense-textured (aphanitic), igneous rock relatively high in iron and magnesia minerals and relatively low in silica, generally dark gray to black, and feldspathic. A general term in contradistinction to felsite, a light-colored feldspathic and highly siliceous rock of similar texture and origin.

Base – The bottom course of a stone wall, or the vertical member above grade of a finished floor.

Bearing Wall – A wall that supports loads, (i.e. floor loading) in addition to its own weight.

Bed – The top or bottom of a joint, natural bed; surface of stone parallel to its stratification.

Bed –

(1) In granites and marbles, a layer or sheet of rock mass that is horizontal, commonly curved and lenticular, as developed by fractures.

(2) In stratified rocks the unit layer formed by sedimentation; of variable thickness, and commonly tilted or distorted by subsequent deformation; generally develops a rock cleavage, parting, or jointing along the planes of stratification.

Bed Joint – A horizontal mortar joint upon which masonry units are placed.

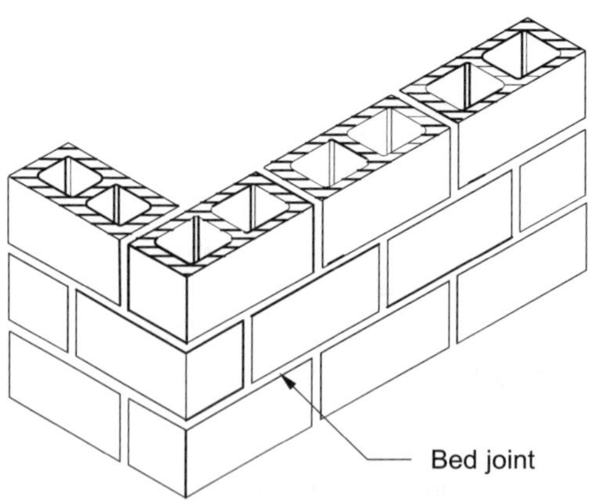

FIGURE 6.4 **Bed mortar joint.**

Belt Course – A continuous horizontal course of flat stones placed in line marking a division in the wall plane. (see ***Sill Course***)

Bevel – When the angle between two sides is greater or less than a right angle.

BIA – Brick Industry Association.

Bluestone – A dense, hard, fine-grained, commonly feldspathic sand-stone or siltstone of medium to dark or bluish-gray color that splits readily along original bedding planes to form thin slabs. Bluestone is not a technical geologic term. It is considered to be a variety of flagstone, with thin, relatively smooth-surfaced slabs being suitable for use as flagging. The term has been applied particularly to sandstones of Devonian age that are being or have been quarried in Eastern New York, Pennsylvania and in Western New Jersey, but similar stones that occur elsewhere may be included. It has also been applied in places to thinly-layered gneisses and schist that can be split and used as flagging, but such stones are not properly embraced by this definition, although they may be marketed properly as flagstone.

Brick – A solid masonry unit of clay or shale, usually formed into a rectangular prism while plastic and burned or fired in a kiln.

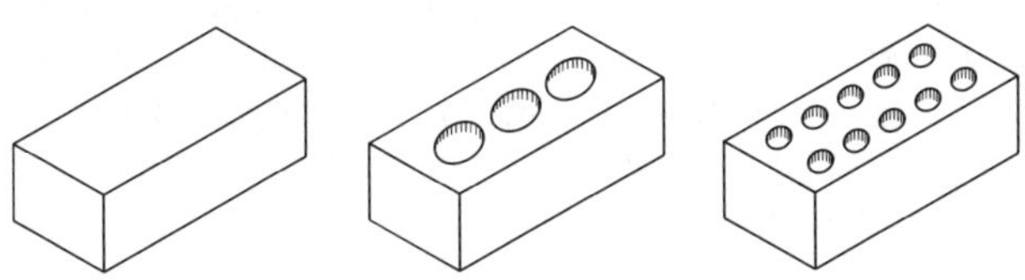

FIGURE 6.5 **Solid masonry units.**

Block – See **Quarry Block**.

Bond Stone – Units used in varying percentages to anchor or bond the stone veneer to the backing material. Bond stones are generally cut to twice the bed thickness of the material being used.

Border Stone – Usually a flat stone used as an edging material. A border stone is generally used to retain the field of the terrace or platform.

Breaking Joints – Any arrangement of masonry units which prevents a continuous vertical joint line.

Broach –

(1) To drill or cut out material left between closely spaced drill holes.

(2) A mason's sharp-pointed chisel for dressing stone.

(3) An inclined piece of masonry filling the triangular space between the base of an octagonal spire and the top of a square tower.

(4) A type of chisel used for working narrow surfaces.

Brownstone – A sandstone of brown or reddish-brown color that is due to a prominent amount of iron oxide.

Brushed Finish – A surface texture obtained by brushing the stone with a coarse rotary-type wire brush.

Building Brick – A brick for building purposes and not specifically treated for appearance such as texture or color. Building bricks conform to ASTM C62. Building bricks are also known as *common bricks*.

Building Stone, Natural – Rock material in its natural state of composition and aggregation as it exists in the quarry and is usable in construction as dimension building stone.

Bull Nose – Convex rounding of a stone member, such as a stair tread or countertop.

Buttering – Placing mortar on a masonry unit with a trowel.

C

Calcarenite – Limestone composed predominantly of caustic sand-size grains of calcite, or rarely aragonite, usually as fragments of shells or other skeletal structures. Some calcarenites contain oolites (small, spherical grains of calcium carbonate that resemble roe) and may be termed oolite limestone. Calcareous sandstones, in which the calcium carbonate is present chiefly as bonding material, are not included in this category.

Calcite Limestone – A limestone containing not more than five percent of magnesium carbonate.

Calcite Streaks – Description of a white or milky-like streak occurring in stone. It is a joint plane usually wider than a glass seam and has been recemented by deposition of calcite in the crack and is structurally sound.

Camber – A slight arching or upward curve of a beam or a soffit.

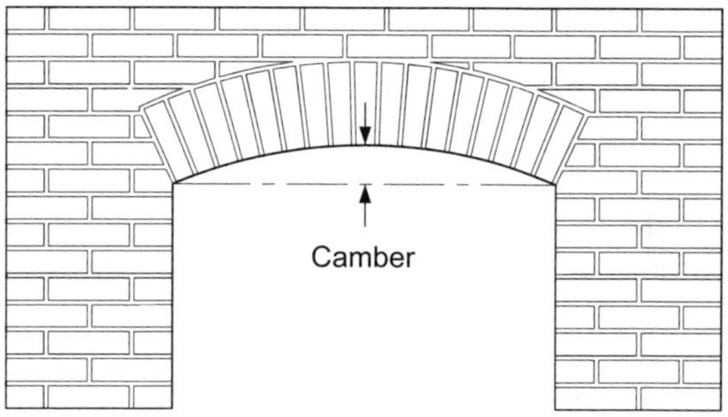

FIGURE 6.6 **Camber.**

Canopy – A sheltering roof, as over a niche or a doorway.

Cantera – A volcanic, quartz-based stone with qualities similar to adoquin, but not as dense, quarried in Mexico.

Capital – The culminating stone at the top of a column or pilaster, often richly carved.

Carve – To decorate by cutting and shaping; to engrave or cut figures as in sculpting.

Caulking – Making a joint tight or leak-proof by sealing with an elastic adhesive compound.

Cavity Vent – A vent or opening in the joints between stones to provide even atmospheric pressure and humidity between the cavity and outside air to prevent condensation and the migration of water into the structure.

Cavity Wall – A wall built of two or more wythes of masonry units arranged to provide a continuous air space within the wall. The wythes are tied together.

Cell – A hollow space within a concrete masonry unit formed by the face shells and the webs having a gross cross-sectional area greater than $1^1/_2$ sq. in. (968 mm^2).

Cement – An ingredient of concrete, mortar and grout. Cement is made by burning a mixture of clay and limestone and then pulverizing into a fine powder.

Cement Putty – A thick, creamy mixture made with pure cement and water which is used to strengthen the bond between the stone and the setting bed. Also called ***Cement Butter, Cement Cream***.

Centering – Temporary formwork for the support of masonry arches or lintels during construction.

Chamfer – To bevel the junction of an exterior angle.

Chat–Sawn Finish – A rough gang saw finish produced by sawing with coarse chat.

Cladding – Non-load-bearing thin stone slabs used for facing buildings.

Clay – A natural mineral aggregate consisting essentially of hydrous aluminum silicate; it is plastic when sufficiently wetted, rigid when dried, and vitrified when fired to a sufficiently high temperature.

Cleavage – The ability of a rock mass to break along natural surfaces; a surface of natural parting.

Cleavage Plane – Plane or planes along which a stone may likely break or delaminate.

Closer – The last masonry unit laid in a course. It may be whole or a portion of unit. Sometimes spelled "closure".

Coating – A protective or decorative covering applied to the surface or impregnated into stone for such purposes as waterproofing, enhancing resistance to weathering, wear, and chemical action, or improving appearance of the stone.

Cobblestone – A natural rounded stone, large enough for use in paving. Commonly used to describe paving blocks, usually granite, generally cut to rectangular shapes.

Collar Joint – The space between two wythes of wall filled with grout.

Commercial Marble – A crystalline rock composed predominantly of one or more of the following materials: calcite dolomite or serpentine, and capable of taking a polish.

Composite – A construction unit in which stone that is to be exposed in the final use is permanently bonded or joined to other material, which may be stone or manufactured material, that will be concealed.

Composite Wall – Multiple wythe construction in which at least one of the wythes is dissimilar to the other wythe or wythes with respect to type or grade of units or mortar.

Concave Joint – A mortar joint formed with a special tool to produce a concave indentation in the mortar joint. This type of mortar joint is weather resistant.

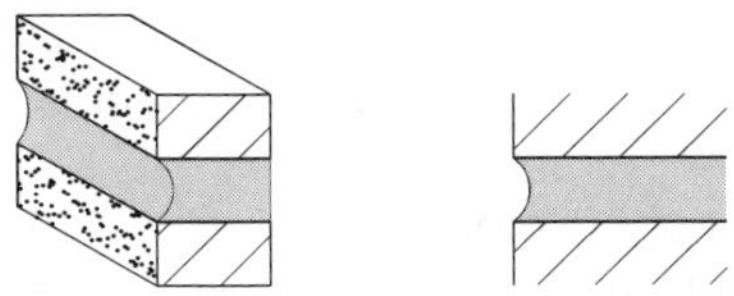

FIGURE 6.7 Concave mortar joint.

Concrete Brick – A solid concrete masonry unit made from portland cement and suitable aggregates with or without additional materials.

Concrete Masonry Unit – Unit made of concrete that consists of natural aggregates that are carefully graded.

Contraction Joints – Spaces where panels meet and which expand as the panels contract.

Control Joint – A groove that is formed, sawed, or tooled in a masonry structure to regulate the location and amount of cracking and separation resulting from the dimensional change of different parts of the structure, thereby avoiding the development of high stresses.

Coping – A flat stone used as a cap on free-standing walls.

Coquina – A limestone composed predominantly of unaltered shells or fragments of shells loosely cemented by calcite. Coquina is generally very coarse-textured and has a high porosity. The term has been applied principally to a very porous shell rock of Eocene age that has been quarried in Florida.

Corbel Plates – Plates of non-ferrous metal fixed into a structure to support stone cladding at intervals and over openings in such a way as not to be visible.

Core – A void space having a gross cross-sectional area less than $1^1/_2$ sq. in. (968 mm^2).

Cornerstone – A stone forming a part of a corner or angle in a wall. Also a stone laid at the formal inauguration of the erection of a building, not necessarily at a corner, usually incorporating a date or inscription.

Cornice – A molded projecting stone at the top of an entablature.

Course – A horizontal range of stone or masonry units the length of the wall.

Coursed Veneer – This is achieved by using stones of the same or approximately the same heights. Horizontal joints run the entire length of the veneered area. Vertical joints are constantly broken so that no two joints will be over one another.

Cross–Bedding – The arrangement of laminations of strata transverse or oblique to the main planes of stratification.

Crowfoot (Styolite) – Description of a dark gray to black zigzag marking occurring in stone. Usually structurally sound.

Crystalline Limestone – A limestone, either calcitic or dolomitic, composed of interlocking crystalline grains of the constituent minerals and of phaneritic texture. Commonly used synonymously with marble and thus representing a re-crystallized limestone. Improperly applied to limestones that display some obviously crystalline grains in a fine-grained mass but which are not of interlocking texture and do not compose the entire mass. (NOTE: All limestones are microscopically, or in part megascopically, crystalline; the term is thus confusing, but should be restricted to stones that are completely crystalline and of megascopic and interlocking texture and that may be classed as marbles).

Culling – Sorting of masonry units (brick and block) for size, color and/or quality.

Cultured Brick – A brick masonry unit made from a material other than clay or shale.

Cultured Stone – Non-naturally occurring stone masonry usually made from concrete.

Curbing – Slabs and blocks of stone bordering streets, walks and similar applications.

Curtain Wall – A non-load bearing wall built between columns and between floor levels to enclose and separate the building from the outside environment.

Cut Joint – Masonry bed joints and head joints cut flush with a trowel.

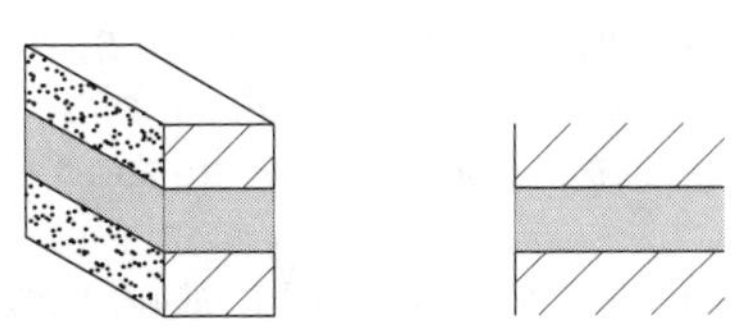

FIGURE 6.8 Cut or flush joint.

Cut Stone – This includes all stone cut or machined to give size, dimension or shape, and produced in accordance with working or shop drawings which have been developed from the architect's structural drawings.

Cutting Stock – A term used to describe slabs of varying size, finish, and thickness which are used in fabricating treads, risers, copings, borders, sills, stools, hearths, mantels, and other special purpose stones.

D

Dacite – A fine-grained, extrusive (volcanic) rock, intermediate in color and composition between basalt and rhyolite.

Damp Course – In stone masonry, an impervious horizontal layer to prevent vertical penetration of water in a wall. May be a course of tile or tight stone (e.g., slate or dense limestone), or a thin layer of asphaltic or bituminous material or metal. Damp courses are generally near grade to prevent upward migration by capillarity action, but they are also used below copings, above roof level in chimneys, and elsewhere to stop downward seepage.

Dampproofing – Prevention of moisture penetration due to capillary action by the addition of one or more coatings of a compound that is impervious to water.

Dentil – Block projections on an entablature.

Dentil Course – The lower part of the cornice with dentils. The cornice is jointed to allow machine production of the dentils.

Dentils – Small, rectangular blocks under a classical cornice, resembling a row of teeth.

Diamond Sawed – Finish produced by sawing with a diamond-toothed circular or gang saw.

Die – A covering layer of interior stone from wall to ceiling.

Dimension Stone – Quarried stones, generally two feet or more square, of a specified thickness, usually with one or more mechanically dressed surfaces.

Dog's Tooth – A brick so laid that its corners project out from the face of the wall.

Dolomitic Limestone – A limestone rich in magnesium carbonate, frequently somewhat crystalline in character. It is found in ledge formations in a wide variety of color tones and textures. Typically, its crushing and tensile strengths are greater than the oolitic limestones and its appearance shows greater variety in texture.

Dowel – A short piece of non-ferrous metal or slate fixed into a mortice or sinking in the joints of adjoining stones to prevent movement.

Dressed or Hand–Dressed – The cutting of rough chunks of stone by hand to create a square or rectangular shape. A stone which is sold as dressed stone generally refers to stone ready for installation.

Drip – A recess cut beneath and slightly behind a projecting stone to prevent water from running down the face of the wall below.

Dripstone – A projecting molding over the heads of doorways, windows and archways to throw off the rain. Also known as a "hoodmold" and, when rectangular, as a "label".

Dry – An open or unhealed joint plane not filled with calcite and not structurally sound.

Dry Pack – A mixture of cement and fine aggregate with only enough moisture for hydration. The material is dry enough so that it must be rammed into place.

Dry Press Brick – Brick formed in molds under high pressure from relatively dry clay, 5% to 7% moisture content.

Dry Wall – A dry wall is a stone wall that is constructed one stone upon the other without the use of any mortar, often used for retaining walls.

Durability – The measure of the ability of natural building stone to endure and to maintain its essential and distinctive characteristics of strength, resistance to decay, and appearance, with relation to a specific manner, purpose, and environment of use.

E

Efflorescence – A crystalline deposit appearing on masonry surfaces typically caused by soluble salts carried through or onto the face of masonry by moisture, which has sometimes been found to come from brick, tile, concrete blocks, cement, mortar, concrete, and similar materials in the wall or above.

Entablature – In classical architecture, the upper part of an order, comprising architrave, frieze, and cornice.

Expansion Bolt – A socket that grips a drilled hole in stone by expanding as the bolt is screwed into it.

Expansion–Contraction Joint – A joint in a wall designed to allow the expansion and contraction of the wall due to temperature change. An expansion joint compresses as panels expand, a contraction joint expands as panels contract.

Exposed Aggregate – Phrase applied to the larger pieces of stone aggregate purposefully exposed for their color and texture in a cast slab.

F

Face – This refers to the exposed portion of stone or masonry unit. The word "face" can also be used when referring to the edge treatment on various cutting stock materials.

Face Brick – A brick made for facing on wall, often treated to produce desired appearance such as surface texture and/or color.

Faced Wall – A wall in which the masonry facing and the backup wall are of different materials and bonded together to act as one wall.

Facing – Any masonry, forming an integral part of a wall, used as a finished surface. See also ***Veneer***.

Fascia – A horizontal belt or vertical face often used in combination with moldings.

Ferruginous – Limestone or sandstone containing a high proportion of iron oxide.

Field Stone – Loose blocks separated from ledges by natural processes and scattered through or upon the regolith ("soil") cover; applied also to similar transported materials, such as glacial boulders and cobbles.

Filling – Filling the natural voids and veins in a stone with material (cement, shellac, or synthetic resins and similar materials often mixed with stone fines).

Fines – The residue resulting from the normal fabrication and processing of stone.

Finish – The final appearance exposed stone slab surfaces are fabricated to meet.

Fireproof – Relatively incombustible.

Fire Brick – Brick made of refractory ceramic material which will resist cracking when exposed to high temperatures.

Flagging – A pavement made of stone slabs is known as flagging or flagstone.

Flagstone – Thin slabs of stone used for paving walks, driveways, patios and similar applications. It is generally fine-grained sandstone, bluestone, quartzite or slate, but thin slabs of other stones may be used.

Flashing – A thin impervious material placed in mortar joints and through air spaces in masonry to prevent water penetration and provide water drainage.

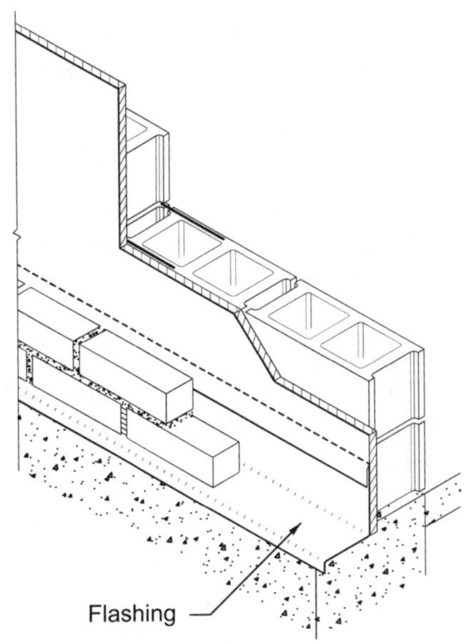

FIGURE 6.9 **Flashing material.**

Fleuri Cut – Cutting quarried marble or stone parallel to the natural bedding plane.

Freestone – A stone that may be cut freely in any direction without fracture or splitting.

Fretwork – Any ornamental openwork or work in relief.

Frieze – A belt course, sometimes decorated with sculpture relief, occurring just under a cornice.

G

Gang Sawed – Description of the granular surface of stone resulting from gang sawing alone.

Gauged or Gauging – A grinding process to make all pieces of material to be used together the same thickness.

Glass Seam – Description of a narrow glass-like streak occurring in stone. It is a joint plane that has been recemented by deposition of translucent calcite in the crack and is structurally sound.

Grade Course – Beginning course at the grade level, generally waterproofed with a dampcheck or damp course.

Grain – The easiest cleavage direction in a stone. "With the grain" same as "natural bed."

Granite – A fine to coarse-grained, igneous rock formed by volcanic action consisting of quartz, feldspar, and mica, with accessory minerals. Granite-type rocks include those of similar texture and origin.

Granite (scientific definition) – A visibly granular, crystalline rock of predominantly interlocking texture, composed essentially of alkalic feldspars and quartz; this is true granite. Feldspar is generally present in excess of quartz, and accessory minerals (chiefly micas, hornblende, or more rarely pyroxene) are commonly present. The alkalic feldspars may be present (1) as individual mineral species, (2) as isomorphous or mechanical inter-growths with each other, or (3) as chemical inter-growths with the lime feldspar molecule, but 80 + 3 percent of the feldspar must be composed of the potash or soda feldspar molecules.

Granite (commercial/building use) – A term that includes granite (as defined above), gneiss, gneissic granite, granite gneiss, and the rock species known to petrologists as syenite, monzonite, and granodiorite, species intermediate between them, the gneissic varieties and gneisses of corresponding mineralogical compositions and the corresponding varieties of porphyritic textures. The term commercial granite shall also include other feldspathic crystalline rocks of similar textures, containing minor amounts of accessory minerals, used for special decorative purposes, and known to petrologists as anorthosite and laurvikite.

Granite Gneiss – A foliated crystalline rock composed essentially of silicate minerals with interlocking and visibly granular texture, and in which the foliation is due primarily to alternating layers, regular or irregular, of contrasting mineralogical composition. In general, a gneiss is characterized by relatively thick layers as compared with a schist. According to their mineralogical compositions, gneisses may correspond to other rocks of crystalline, visibly granular, interlocking texture, such as those included under the definition of commercial granite, and may then be known as granite gneiss if strongly foliated, or gneissic granite if weakly foliated.

Black Granite – Rock species known to petrologists as diabase, diorite, gabbro, and intermediate varieties are sometimes quarried as building stone, chiefly for ornamental use, and sold as "black granite". As dimension blocks or slabs, they are valued specifically for their dark gray to black color when polished. Scientifically, they are far removed in composition from true granites though they may be satisfactorily used for some of the purposes to which commercial granites are adapted. They possess an interlocking crystalline texture, but unlike granites, they contain little or no quartz or alkalic feldspar, and are characterized by an abundance of one or more of the common black rock-forming minerals (chiefly pyroxenes, hornblende, and biotite).

Greenstone – Includes stones that have been metamorphosed or otherwise changed and have assumed a distinctive greenish color owing to the presence of one or more of the following minerals: chlorite, epidote, or actinolite.

Grounds – Nailing strips placed in masonry walls as a means of attaching trim or furring.

Grout – Pourable cementitious material.

Coarse Grout – Used for wide grout spaces 2 in. (50.8 mm) or more, consisting of one part Portland cement, two-and-a-quarter to three parts sand, and one to two parts pea gravel.

Fine Grout – Used in narrow grout spaces, consisting of one part Portland cement and two-and-a-quarter to three parts sand.

H

Hacking – Laying masonry units so that the bottom edge is set back from the plane surface of the wall.

Hand–Cut Random Rectangular Ashlar – A pattern where all the stone is hand cut into squares and rectangulars. Joints are fairly consistent. Similar to sawed-bed ashlar in appearance.

Hand or Machine Pitch–Faced (Rock–Faced) Ashlar – A finish given to both veneer stone and cutting stock. This is created by establishing a straight line back from the irregular face of the stone. Proper tools are then used to cut along the line, leaving a straight arris and the intended rustic finish on the face.

Harsh Mortar – A mortar that is difficult to spread. it is not workable.

Head – The end of a stone which has been tooled to match the face of the stone. Heads are used at outside corners, windows, door jams, or any place where the veneering will be visible from the side.

Head Joint – The vertical mortar joint between ends of masonry units. Sometimes called a cross joint.

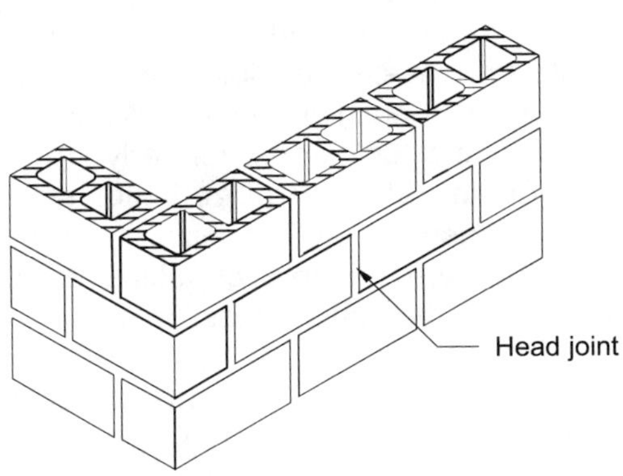

FIGURE 6.10 **Head joint.**

Hearth Stone – Originally the single large stone or stones used for the hearth, now most commonly used to describe the stone in front of the fire chamber and many times extending on either or both sides of the front of the fire chamber.

Header – A masonry unit which overlaps two or more adjacent wythes of masonry to tie them together, also called a bonder.

High Lift Grouting – The technique of grouting masonry in lifts for the full height of the wall.

Holes – Sinkages in the top beds of stones to engage Lewis pins for hoisting.

Honed Finish – Honed is a fine smooth finish, though not as fine as a polished finish.

Hollow Brick – A brick with interior voids creating bearing surface of less than 75% of the gross cross-sectional area.

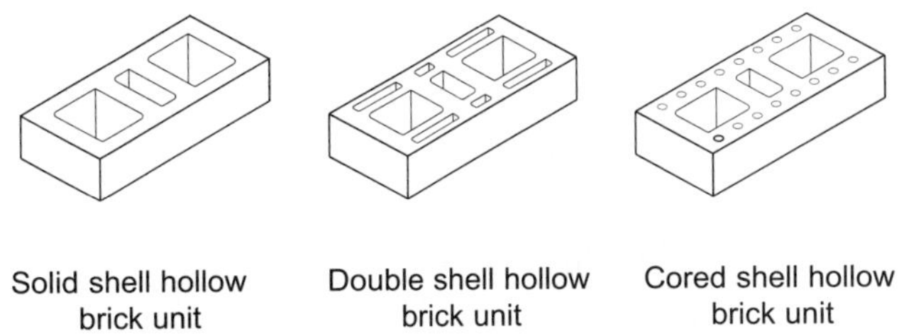

FIGURE 6.11 **Hollow brick units.**

I

IBC – International Building Code.

ICC – International Code Council.

Igneous – One of the three classes of rock (igneous, sedimentary and metamorphic), solidified from molten state, as granite and lavas.

Impervious – The quality of resisting moisture penetration.

Impregnation – Applying a chemical containing stain inhibitors that penetrates below the surface of the stone.

Incise – To cut inwardly or engrave, as in an inscription.

Inscription – Lettering cut in stone.

In Situ – A Latin phrase meaning – in the original or natural place. It is used in construction to reference what was actually built in the field.

Initial Rate of Absorption – The amount of water a dry brick will absorb in one minute. The *initial rate of absorption* is not to be confused with *absorption*. There is no consistent relationship between the *initial rate of absorption* and *absorption*.

J

Joint – The space between stone units, usually filled with mortar.

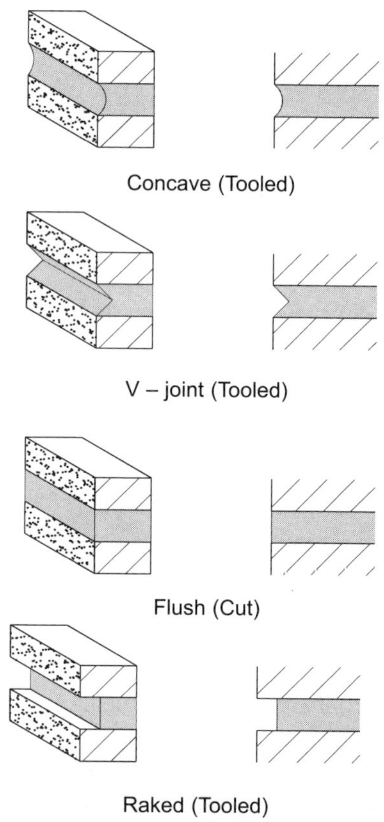

FIGURE 6.12 **Mortar type joints.**

Jointer – A tool used by the bricklayers to form various types of mortar joints, such as concave or "V".

Jointing – The process of finishing a mortar joint into a specific shape, such as concave or "V".

Jointing Scheme – A detailed architectural drawing showing the dimensions, locations and configurations of stone units and joints on the structure.

Jumper – In ashlar patterns, a piece of stone of higher rise than adjacent stones which is used to end a horizontal mortar joint at the point where it is set.

K

Kerf – A slot cut into the edge of a stone with a saw blade for insertion of anchors.

L

Lava – A general term applied to igneous rocks, such as basalt and rhyolite, that erupted from the earth by volcanic action.

Lead Buttons – Lead spacers in the solid horizontal joints to support the top stones until the mortar has set.

Lean Mortar – Mortar containing a low amount of cementitious material.

Lewis Holes – Holes in cut stones for lifting and support during setting of cut stones and sometimes for permanent support. Holes are checked for the particular Lewis lifting device or hook to be used.

Lifts – The vertical section of grout placed in a masonry wall in a single pass.

Limestone – A sedimentary rock composed of calcium carbonate; includes many varieties. (See oolitic limestone, dolomitic limestone, crystalline limestone.) Limestones that contain not more than five percent magnesium carbonate may be termed calcite limestone, as distinguished from those that contain between five and 40 percent magnesium carbonate (magnesium or dolomitic limestone), and from those that contain in excess of 40 percent as the mineral dolomite (dolostone, formerly known as the rock dolomite). Recrystallized limestones and compact, dense, relatively pure microcrystalline varieties that are capable of taking a polish are included in commercial marbles.

Liners – Strengthening elements attached to the back of stone slabs, usually a structurally sound section of similar stone doweled and epoxied into place.

Lintel – The block of stone spanning the top of an opening such as a door-way or window; sometimes called a head.

Lipping – Usually refers to paving applications. Lipping is caused when two pieces of material to be joined together are slightly warped or twisted causing one or more edges to be higher or lower than the adjoining material.

Low Lift Grouting – The technique of grouting as the wall is constructed.

Lug Sill – A stone sill set into the jambs on each side of masonry opening.

M

Machine Finish – The generally recognized standard machine finish produced by planers.

Malpais – Literally, badland; refers to dark-colored rock, commonly lava, in rough terrain. As defined for architectural use: calcium carbonate with other components which give it color, markings, and texture suitable as a desirable building stone.

Marble – A metamorphic limestone in a more or less crystalline state capable of taking a high polish. Occurs in a wide range of colors and variations. Marble that contains less than five percent magnesium carbonate may be termed calcite marble; from 5 to 40 percent magnesium carbonate, magnesian or dolomitic marble; and more than 40 percent dolomite marble. These limiting values are, however, not strictly established in petrologic science and are used herein as arbitrary limits.

> ***Onyx*** – So called in trade, is a crystalline form, commonly microcrystalline, of calcium carbonate deposited usually from cold-water solutions. It is generally translucent and shows a characteristic layering. The term onyx marble is technically a misnomer, as true onyx is a variety of cryptocrystalline fibrous silica (chalcedony), and is closely related in form and origin to agate.
>
> ***Serpentine*** – A marble characterized by a prominent amount of the mineral serpentine.
>
> ***Travertine*** – A form of limestone precipitated from ground waters, as in caves or in orifices of springs (see Limestone).
>
> ***Verde Antique*** – A commercial marble composed chiefly of massive serpentine and capable of taking a high degree of polish. Verde antique is not a true marble in the scientific sense, but is commonly sold as a decorative commercial marble and requires the adjectival modifier verde (or verd) antique. Verde antique is commonly veined with carbonate minerals, chiefly calcite and dolomite.

Mason – A worker skilled in laying brick, block or stone; as in stonemason or brickmason.

Masonry – Built up construction, usually a combination of materials with units set in mortar.

Masonry Unit – Natural or manufactured building unit of burned clay, concrete, stone, glass, gypsum or other similar material.

Mechanical Bond – Tying masonry units together with metal ties or reinforcing steel.

Metamorphism – The change or alteration in a rock caused by exterior agencies, such as deep-seated heat and pressure, or intrusion of rock materials.

MIA – Masonry Institute of America.

Miter Joint – The junction of two units at an angle of which the junction line usually bisects on a 45° angle.

Modular Multiple–Cut (Pattern–Cut) – This refers to standard patterns used throughout the stone industry. These patterns are usually based on multiples of a given height. Stone that is multiple cut or pattern cut is pre-cut to allow typically for $^1/_4$ in. (6.4 mm) or $^1/_2$ in. (12.7 mm) joints or beds.

Moldings – Decorative stone deviating from a plane surface by projections, curved profiles, recesses or any combination thereof.

Mortar – A plastic mixture of cement, lime, sand and water used to bond masonry units.

Mosaic – A veneering which is generally irregular with no definite pattern. Nearly all the stone used in a mosaic pattern is irregular in shape.

Multi-Unit Wall – A masonry wall composed of two or more wythes.

N

NCMA – National Concrete Masonry Association.

Natural Bed – The setting of stone on the same plane as it was formed in the ground. This typically applies to all stratified materials.

Natural Cleft – This generally pertains to stones which are formed in layers in the ground. When such stones are cleaved or separated along a natural seam the remaining surface is referred to as a natural cleft surface.

Natural Stone – A product of nature. A stone such as granite, marble, limestone, slate, travertine or sandstone that is formed by nature and is not artificial or manmade.

Neat Cement – A pure cement uncut by a sand admixture.

Nicked Bit Finish – Obtained by planing the stone with a planer tool in which irregular nicks have been made in the cutting edge.

Non-Staining Mortar – Mortar composed of materials which individually or collectively do not contain material that will stain. Usually having a very low alkali content.

Nominal Dimension – A dimension greater than the actual masonry unit dimension by the thickness of a mortar joint.

Notch – A V–shaped cut made on the edge or head of a stone.

O

Obsidian – A glassy phase of lava.

Onyx Marble – A dense, crystalline form of lime carbonate usually deposited from cold-water solutions. Generally translucent and shows a characteristic layering due to mode of accumulation.

Oolitic Limestone – A calcite-cemented calcareous stone formed of shells and shell fragments, practically non-crystalline in character. It is found in massive deposits located almost entirely in Lawrence, Monroe, and Owen Counties, Indiana and in Alabama, Kansas, and Texas. This limestone is characteristically a freestone, without cleavage planes, possessing a remarkable uniformity of composition, texture and structure. It possesses a high internal elasticity, adapting itself without damage to extreme temperature changes.

Opalized – The introduction into a rock of siliceous material in the form of opal, hydrous silicate.

Ornamental Facing – A design formed by the laying of stone, brick, tile or other masonry units so to produce a decorative effect.

Out of Wind – To be out of wind is to have the arris of the stone not in parallel or perpendicular lines. Stone which is out of wind has an irregular or rustic appearance.

P

Palletized – A system of stacking stone on wooden pallets. Stone which comes palletized is easily moved and transported by modern handling equipment. Palletized stone will arrive at the job site in better condition than unpalletized material.

Parapet Wall – That part of any wall entirely above the roof line.

Parging – Plastering a cementitious coating of mortar onto a surface, often used for dampproofing.

Parquetry – An inlay of stone floors in geometrical or other patterns.

Perforated Wall – A wall containing a considerable number of relatively small openings. Often called pierced wall or screen wall.

Permeability – The quality of allowing the passage of fluids.

Perrons – Slabs of stone set on other stones serving as steps and arches in gardens.

Phenocryst – In igneous rocks, the relatively large and conspicuous crystals in a finer-grained matrix or ground mass.

Pick and Dip – A method of laying brick whereby the mason simultaneously picks up a brick with one hand and, with the other hand, picks up enough mortar on the trowel to lay the brick. Also known as the *Eastern or New England Method.*

Pilaster – An engaged pier of shallow depth; in classical architecture it follows the height and width of related columns, with similar base and cap.

Pitched Stone – Stone having arris clearly defined; face, however, is roughly cut with pitching chisel used along the line which becomes the arris.

Plasticizers – An additive to mortar or grout to impart more plastic quality.

Plinths – The lower square part of the base of a column. A square base or a lower block, as of a pedestal. The base block at the juncture of baseboard and trim around an opening.

Plucked Finish – Obtained by rough-planing the surface of stone, breaking or plucking out small particles to give rough texture.

Pointing – The filling and tooling of mortar joints with mortar or caulking compounds.

Polished – The finest and smoothest finish available in stone characterized by a gloss or reflective property. Generally performed only on hard, dense materials.

Porphyry – An igneous rock in which relatively large and conspicuous crystals (phenocrysts) are set in a matrix of finer crystals.

Pressure Relieving Joint – An open horizontal joint below the supporting angle or hanger located at approximately every floor line and not over 15 ft (4.5 m) apart horizontally and every 20 to 30 ft (6.1 m to 9.1 m) vertically to prevent the weight from being transmitted to the masonry below. These joints are to be caulked with a resilient non-staining material to prevent moisture penetration.

Projections – This refers to the pulling out of stones in a wall to give an effect of ruggedness. The amount a stone can be pulled out can vary between $^1/_2$ in. and $1^1/_2$ in. (12.7 mm and 38.1 mm). Stones are either pulled out at the same degree at both ends or sometimes one end is pulled out, leaving the other end flush with the majority of veneer.

Pumice – An exceptionally cellular, glassy lava resembling a solid froth.

Q

Quarry – An excavation where usable stone is extracted from the ground.

Quarry Block – A rectangular piece of rough stone as it comes from a quarry, frequently dressed (scabbed) or wire–sawed for shipment.

Quartz – A silicon dioxide mineral that occurs in colorless and transparent or colored hexagonal crystals and also in crystalline masses. One of the most common minerals of sandstone.

Quartz-Based Stone – A stone that may be either sedimentary in formation (as in sandstone, or methamorphic, as in quartzite).

Quartzite – A compact granular rock composed of quartz crystals, usually firmly cemented to make the mass homogeneous. The stone is generally quarried in stratified layers, the surfaces of which are unusually smooth. Its crushing and tensile strengths are extremely high; the color range is wide.

Quartzitic Sandstone – A sandstone with a high concentration of quartz grains and siliceous cement.

Quirt – A groove separating a bed or other molding from the adjoining members.

Quirk–Miter – Linear edge work for corner joints.

Quoins – Stones at the corner of a wall emphasized by size, projection, rustification, or by a different finish.

R

Rabbet – A groove cut into the surface along an edge to receive another piece similarly cut.

Raggle – A groove in a joint or a special unit to receive roofing or flashing.

Rake – An angular cut on the face of a stone.

Raked Joint – A type of mortar of joint that has the mortar raked out of it to a specified depth, before the mortar has set.

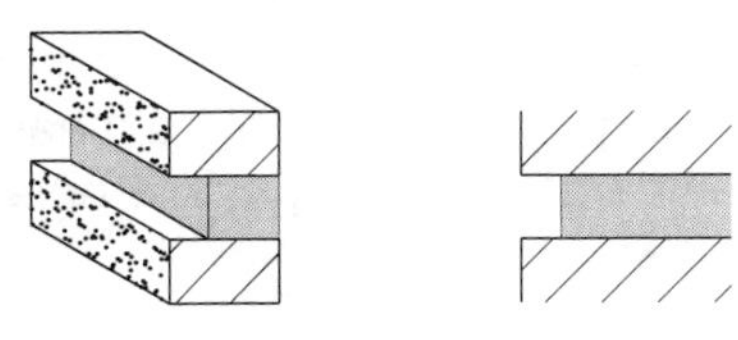

FIGURE 6.13 **Raked (tooled) joint.**

Range – A course of any thickness that is continued across the entire face. All range courses need not be of the same thickness.

Recess – A sinkage in a wall plane.

Reglet – A recess used to receive and secure flashing.

Relief or Relieve – Ornament in relief. The ornament or figure can be slightly, half, or greatly projected.

Retempering – To moisten and remix, to proper consistency, mortar for use. Also known as *Retemper*.

Return – The right angle turn of a molding.

Return Head – Stone facing with the finish appearing on both the face and the edge of the same stone, as on the corner of a building.

Reveal – The depth of stone between its outer face and a window or door set in an opening.

Rift – The most pronounced (see ***Grain***) direction of splitting or cleavage of a stone. Rift and grain may be obscure, as in some granites, but are important in both quarrying and processing stone.

Riprap – Irregular shaped stones used for facing bridge abutments and fills. Stones thrown together without order to form a foundation or sustaining walls.

Rise – The word "rise" refers to the heights of stone. Generally used in reference to veneer stone.

Riser – The vertical element in a set of stairs, forming the space between one step and the next.

Rock – An integral part of the earth's crust composed of aggregate of grains of one or more minerals. (Stone is the commercial term applied to quarry products.)

Rock (Pitch) Face – This is similar to split face, except that the face of the stone is pitched to a given line and plane producing a bold appearance rather than the comparatively straight face obtained in split face.

Rodding – Reinforcement of a structurally unsound marble by cementing reinforcing rods into grooves or channels cut into the back of the slab.

Rose Window – A circular stone window fitted with carved tracery.

Rough Sawn – A marble surface finish accomplished by the gang sawing process.

Rubbed Finish – Mechanically rubbed for smoother finish.

Rubble – A product term applied to dimension stone used for building purposes, and consisting of irregularly shaped pieces, partly trimmed or squared, generally with one split or finished face, and selected and specified with a size range.

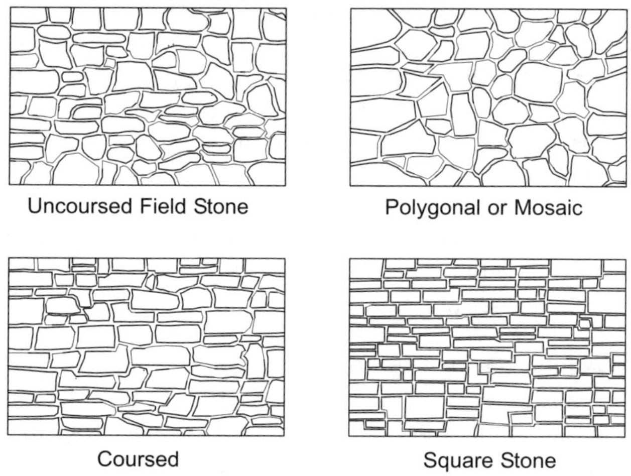

FIGURE 6.14 **Rubble stone.**

Running Bond – Lapping of units in successive courses such that the vertical head joints of alternating course line up.

Rustication – Chamfers or square sinkings around the face edges of individual stones to create shadows and to give an appearance of greater weight to the lower part of a building. When only the horizontal joints are sunk, the device is known as banded rustication.

Rustification – Recessing the margin of cut stone so that when placed together a channel is formed at each joint.

S

Sack Joint – A mortar joint that has been wiped or rubbed with a rag or object such as a rubber heel.

Saddle – A flat strip of stone projecting above the floor between the jambs of a door; a threshold.

Salient – A prominent feature.

Sandblasted – A dull non-glossy finish applied to stone; usually accomplished by blasting air blended with sand across the surface.

Sand–Sawn Finish – The surface left as the stone comes from the gang saw. Moderately smooth, granular surface varying with the texture and grade of stone.

Sandstone – A sedimentary rock consisting of quartz cemented with silica, iron oxide or calcium carbonate. Sandstone is durable, has a very high crushing and tensile strength, and a wide range of colors and textures.

Varieties of sandstone are commonly designated by the kind and prominence of interstitial and bonding materials, as siliceous sandstone (bonding material primarily silica), calcareous sandstone (calcium carbonate prominent as bonding material or as accessory grains or both), argillaceous sandstone (clay minerals prominent as interstitial or bonding materials, or as thin laminac), ferruginous sandstone (iron oxide or hydroxide minerals, or as thin laminac), ferruginous sandstone (iron oxide or hydroxide minerals (hematic, limonite, et. al.) as interstitial or as bonding materials in sufficient amount to impart appreciable color to the stone): brownstone (ferruginous sandstone of dark brown or reddish brown color), arkose, arkosic sandstone, or feldspathic sandstone (a sandstone that contains an abundance of grains of feldspar), conglomerate (a sandstone composed in large part of rounded pebbles, also called puddingstone).

The term "brownstone" was applied originally to certain Trassic sandstones of the Connecticut Valley in Massachusetts (Longmeadow sandstone), Connecticut (Portland sandstone), and to similarly appearing reddish-brown sandstone quarried in and near Hummelstown, Pennsylvania. Thus the term originally had geographic significance, but such geographic limitation is undesirable.

Sawed Edge – A clean cut edge achieved by cutting with a diamond blade, gang saw or wire saw.

Sawed Face – A finish obtained from the process used in producing building stone. Varies in texture from smooth to rough depending on the type of materials used in sawing; characterized as diamond sawn, sand sawn, chat sawn, and shot sawn.

Scale – Thin lamina or paper-like sheets of rock, often loose, and interrupting an otherwise smooth surface on stone.

Schist – A loose term applying to foliated metamorphic rock (recrystallized) characterized by thin foliate that are composed predominantly of minerals of thin platy or prismatic habits and whose long dimensions are oriented in approximately parallel positions along the planes of foliation. Because of this foliated structure schists split readily along these planes and possess a pronounced rock cleavage. The more common schists are composed of the micas and other mica-like minerals (such as chlorite) and may contain subordinate quartz and/or feldspar of comparatively fine-grained texture.

Scoria – Irregular masses of lava resembling clinker of slag; may be cellular (vesicular) dark-colored and heavy.

Scotia – A concave molding.

Sculpture – Statuary cut from stone by a sculptor using hand tools and polishing materials.

Scutch – A mason's cutting tool used for dressing and trimming brick to a desired shape. It resembles a small pick and is sometimes call a "scotch".

Sealant – An elastic adhesive compound used to seal stone veneer joints.

Semi–Rubbed – A finish achieved by rubbing, by hand or machine, the rough or high spots off the surface to be used leaving a certain amount of the natural surface along with the smoothed areas.

Serpentine – A hydrous magnesium silicate material of igneous origin, typically a very dark green color with markings of white, light green or black. One of the hardest varieties of natural building stone.

Set – To change from a plastic to a hard state.

Setter – An experienced journeyman who installs dimension stone.

Setting – The trade of installing dimension stone.

Setting Space – A term used to indicate the distance from the finished face of the marble to the face of the backup wall.

Shear – A type of stress; a body is in shear when it is subjected to a pair of equal forces which are opposite in direction and which act along parallel planes.

Shot–Sawn – Description of a finish obtained by using steel shot in the gang sawing process to produce random markings for a rough surface texture.

Shot–Sawn Finish – A rough gang-saw finish produced by sawing with chilled steel shot.

Sill – A flat stone used under windows, doors, and other masonry openings.

Sill Course – A narrow horizontal course of masonry, sometimes slightly projected such as window sills which are made continuous.

Siltstone – A fine-grained non-carbonate clastic rock composed of at least 67 percent of detrital grains of quartz and silicate minerals of silt size. Siltstones are rarely marketed as such but commonly are considered as fine-grained sandstones. This class of sediments is texturally transitional between sandstones and shales (mudstones). Many bluestones and siliceous flagstones fall within this category. The term is included in these definitions to explain the relationship of some siliceous flagstones to the sandstone category.

Slab – A lengthwise cut of a large quarry block of stone.

Slate – A very fine-grained metamorphic rock derived from sedimentary rock shale. Characterized by an excellent parallel cleavage entirely independent of original bedding, by which the rock may be split easily into relatively thin slabs.

Essential mineral constituents of slates are usually members of the mica group, commonly sericite, muscovite, and paragonite; of the clay group, illite and kaolinite; and of the chlorite group. Common accessory minerals are iron oxides, calcite, quartz, and feldspar. Other minerals may be present also as minor accessories. Most slates are derived from shales. Others are derived from fine-grained igneous rock, chiefly volcanic tuffs, but these are rare and of little commercial importance.

Slip Sill – A stone sill set between the jambs. (Also called ***Lug Sill***).

Slurry – A thin watery mixture of neat cement, or cement and sand.

Slurry Coat – A brushed application of slurry, applied to the back of adhered veneer units and the support backing.

Smooth Finish – Description of the finish produced by planer machines plus the removal of objectionable tool marks. Also known as "smooth planer finish" and "smooth machine finish."

Snapped Edge, Quarry Cut or Broken Edge – This refers to a natural breaking of a stone either by hand or machine. The break should be at right angles to the top and bottom surface.

Soapstone – A soft metamorphic rock composed of the mineral talc, used for hearths, washtubs, table tops, and carved ornaments.

Soffit – The finished, exposed underside of a lintel, arch, or portico.

Solid Brick – A brick with interior voids creating bearing surfacing that are at least 75% of the gross cross-sectional area. The term solid brick does not necessarily mean 100% solid.

Spall – A stone fragment that has split or broken off.

Spandrel Wall – That part of a curtain wall above the top of a window in one story and below the sill of the window in the story above.

Splay – A beveled or slanted surface.

Spline – A thin strip of material, such as wood or metal, inserted into the edges of two stone pieces or stone tiles to make a butt joint between them.

Split – Division of a rock by cleavage.

Split Face (Sawed Bed) – Usually split face is sawed on the beds and is split either by hand or with machines so that the surface face of the stone exhibits the natural quarry texture.

Splitstone Finish – Obtained by sawing to accurate heights then breaking by machine to required bed widths. (Normal bed widths are $3^1/_2$ in. (88.9 mm)).

Spot or Spotting – An adhesive contact, usually of plaster of paris, applied between the back of marble veneer and the face of the backup wall to plumb or secure standing marble.

Stacked Bond – Stone that is cut to one dimension and installed with unbroken vertical and horizontal joints running the entire length and height of the veneered area.

Statue – A three-dimensional form or likeness sculpted, modeled, carved, or cast in material such as stone, clay, wood or bronze.

Sticking – Trade language for connecting pieces of stone together by using cement or epoxies.

Stone – Sometimes synonymous with rock, but more properly applied to individual blocks, masses or fragments taken from their original formation or considered for commercial use.

Stool – A flat stone, generally polished, used as an interior sill.

Stratification – A structure produced by deposition of sediments in beds or layers (strata), laminae, lenses, wedges, and other essentially tabular units.

Strip Rubble – Strip rubble comes from a ledge quarry. The beds of the stone, while uniformly straight, are of the natural cleft as the stone is removed from the ledge, and then split by machine to approximately four inch widths.

Strips – Long pieces of stone, usually low height ashlar courses, where length to height ratio is at maximum for the material used.

Struck Joint – A mortar joint which is formed with a recess at the bottom of the joint.

Styolite – A longitudinally streaked, columnar structure occurring in some marbles and of the same material as the marble in which it occurs.

T

Tablet – A small, flat slab or surface of stone, especially one bearing or intended to bear an inscription, carving or the like.

Temper – *See Retempering.*

Template – A pattern for repetitive marking or fabricating operation.

Tender – A laborer who tends to the needs of the mason. Also know as *hodcarrier.*

Terra Cotta – Non-structural hard-burned, glazed, or unglazed clay building units usually used for ornamental purposes.

Terrazo – A type of concrete in which chips or pieces of stone, usually marble, are mixed with cement and are ground to a flat surface, exposing the chips, which take a high polish.

Texture – Surface quality of stone independent of color.

Textured Finish – A rough surface finish.

Thin Brick – Brick specifically designated for adhered veneer application. Common thickness are 1/2 in. to 1 in. with typical maximum thickness of 1 3/4 in.

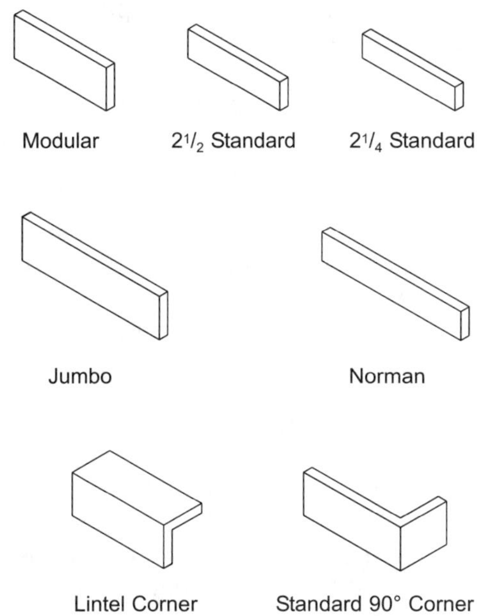

FIGURE 6.15 Types of thin brick units.

Tie – Any unit of material which connects masonry to masonry or other materials.

Thin Stone – Stone slabs generally of 2 in. (50.8 mm) or less in thickness.

FIGURE 6.16 Thin stone units.

Tolerance – Specified allowance of variation from installation location.

Tooling – Compressing and shaping the face of a mortar joint with a special tool other than a trowel. Also known as ***Jointing***.

Toother – A brick projecting from the end of a wall against which another wall will be built.

Tracery – Ornamentation of panels, circular windows, window heads, etc.

Translucence – Permitting light to pass through with little diffusing. Certain marble varieties are translucent.

Travertine Limestone – A variety of limestone that has a partly crystalline or microcrystalline texture and porous or cellular layered structure, the cells being usually concentrated along certain layers and commonly displaying small stalactitic forms.

Travertine Marble – A variety of limestone regarded as a product of chemical precipitation from hot springs. Travertine is cellular with the cells usually concentrated in thin layers that display a stalactitic structure. Some that take a polish are sold as marble and may be classified as travertine marble under the class of "Commercial Marble."

Tread – A flat stone used as the top walking surface on steps.

Trim – Stone used as decorative items only, such as sills, coping and similar applications with the facing of another material.

Trimmer Arch – A stone arch, usually a low-rise arch, used for supporting a fireplace hearth.

Tuff – Cemented volcanic ash.

U

Undercut – Cut to project an overhanging part.

Unit – A piece of fabricated cubic or thin dimension stone.

V

V Joint – Tooling that works the mortar tight and provides a good weather joint. Used to emphasize joints and conceal small irregularities in laying and provide a line in center of mortar joint.

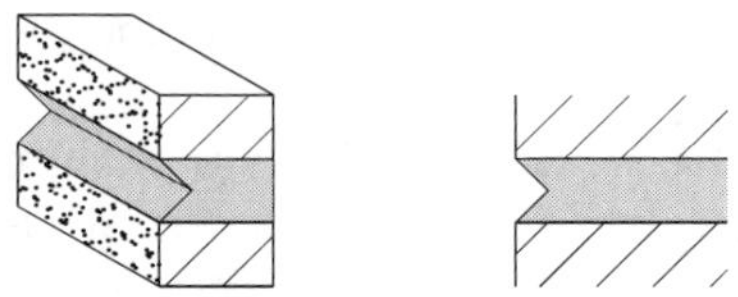

FIGURE 6.17 V mortar joint.

Vein Cut – Cutting quarried marble or stone perpendicular to the natural bedding plane.

Veinings – Colored markings in limestone, marble, alabaster and other stone.

Veneer – A single wythe of masonry for facing purposes, not considered to contribute to the structural integrity of the wall.

Veneer Stone – Any stone used as a decorative facing material.

Venting – Creating an outlet in a wall for air and moisture to pass through. (see ***Cavity Vent***).

Verd (or Verde) Antique – A marble suitable for exterior usage composed chiefly of massive serpentine and capable of being polished. It is commonly crossed by veinlets of other minerals, chiefly carbonates of calcium and magnesium.

Vitrification – The condition resulting when kiln temperatures are sufficient to fuse grains of clay and to close pores of a clay product rendering the product impervious.

Vitrified – To be fused together by heat.

Vug – A cavity in rock sometimes lined or filled with either amorphous or crystalline material, common in calcareous rocks such as marble or limestone.

W

Wall (Embed) Plate – A horizontal member anchored to a masonry wall to which other structural elements may be attached.

Walls –

(1) ***Bearing***: A wall supporting a vertical load in addition to its own weight.

(2) ***Cavity***: A wall in which the inner and outer wythes are separated by an air space, but tied together with metal ties.

(3) ***Composite***: A wall in which the facing and backing are of different materials and bonded together to exert a common reaction under load.

(4) ***Veneer or Faced***: A wall in which a thin facing and the backing are of different materials, but not bonded to exert a common reaction under load.

(5) ***Wind (Wined)***: A twisting warp from cutting slabs in the gang saws.

(6) ***Wythe***: A continuous vertical section of wall one masonry unit in thickness.

Wall Tie – A bonder or metal piece that connects wythes of masonry to each other or to other materials.

Cavity Wall Tie – A rigid, corrosion-resistant metal tie that bonds two wythes of a cavity wall.

Veneer Wall Tie – A strip or piece of metal or wire used to tie a facing to the backing.

Warped Walls – A condition experienced in flagging or flagstone materials; common with flagstone materials that are taken from the ground and used in their natural state. To eliminate warping in stones it would be necessary to further finish the material, such as machining, sand rubbing, honing, or polishing.

Wash – A sloped area or the area water will run over.

Water Bar – Typically a strip in a reglet in window sill and stone below to prevent water passage.

Water–Jet Finish – A surface treatment performed by using water under extreme high pressure.

Water Table – A projection of lower masonry on the outside of the wall slightly above the ground. Often a damp course is placed at the level of the water table to prevent upward penetration of ground water. May require waterproofing on back of stone to eliminate staining and wicking.

Waxing – An expression used in the marble finishing trade to indicate the filling of natural voids with color-blended materials.

Wear – The removal of material or impairment of surface finish through friction or impact use.

Weathering –

(1) Natural alteration by either chemical or mechanical processes due to the action of constituents of the atmosphere, surface waters, soil and other ground waters, or to temperature changes.

(2) The inclined top surface of a stone such as a coping, cornice, or window sill.

Wedging – Splitting of stone by driving wedges into planes of weakness.

Weep Holes – Openings placed in mortar joints of facing material at the level of flashing to permit the escape of moisture.

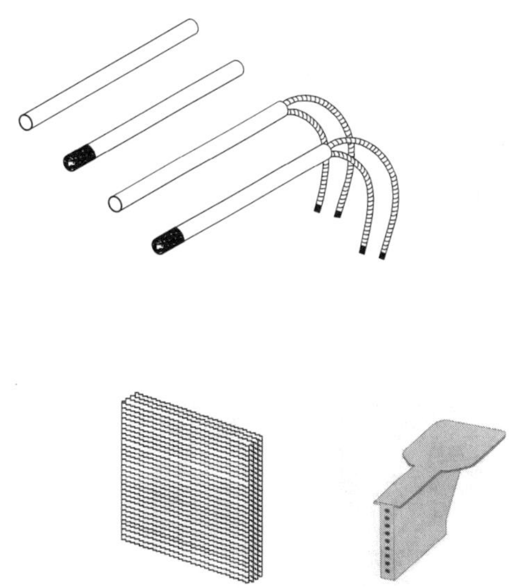

FIGURE 6.18 **Weep hole devices.**

Wire Cut Brick – A brick formed by extruding a plastic clay into a bar shape and wire cutting the extrusion into brick units, before firing.

Wire Saw – A method of cutting stone by passing a twisted, multistrand wire over the stone and immersing the wire in a slurry of abrasive material.

7 REFERENCES

2007 California Building Code, published by International Code Council, 500 New Jersey Avenue, NW 6th Floor, Washington, DC 20001-2070 www.iccsafe.org.

2006 International Building Code, published by International Code Council, 500 New Jersey Avenue, NW 6th Floor, Washington, DC 20001-2070 www.iccsafe.org.

2005 Masonry Standard Joint Committee, *Code, Specification and Commentaries,* published by ACI International, SEI – Structural Engineering Institute of the American Society of Civil Engineers, and TMS – The Masonry Society.

ASTM C55 Standard Specification for Concrete Building Brick.

ASTM C62 Standard Specification for Building Brick (Solid Masonry Units Made From Clay or Shale).

ASTM C90 Standard Specification for Loadbearing Concrete Masonry Units.

ASTM C126 Standard Specification for Ceramic Glazed Structural Clay Facing Tile, Facing Brick, and Solid Masonry Units.

ASTM C144 Standard Specification for Aggregate for Masonry Mortar.

ASTM C216 Standard Specification for Facing Brick (Solid Masonry Units Made From Clay or Shale).

ASTM C270 Standard Specification for Mortar for Unit Masonry.

ASTM C652 Standard Specification for Hollow Brick (Hollow Masonry Units Made From Clay or Shale).

ASTM C1088 Standard Specification for Thin Veneer Brick Units Made From Clay or Shale.

Amrhein, James E; Masonry Veneer, Second Edition. Masonry Institute of America. Los Angeles, CA, 1994. www.masonryinstitute.org.

Architectural Terra Cotta Institute (1961), Public Works Specifications: Ceramic Tile.

Beall, Christine, (June, 1991), Baked Earth, Alexandria, Virginia: The Construction Specifier.

Brandow, G. E., Hart, G. C. and Virdee, A (1993), Design of Masonry Structures, Citrus Heights, California: Concrete Masonry Association of California & Nevada.

BIA *Technical Notes 28,* Anchored Brick Veneer, Wood Frame Construction, www.bia.org.

BIA *Technical Notes 28A,* Brick Veneer Existing Construction, www.bia.org.

BIA *Technical Notes 28B,* Brick Veneer/Steel Stud Walls, www.bia.org.

BIA *Technical Notes 28C,* Thin Brick Veneer–Introduction, www.bia.org.

Brick Industry Association, BIA, 1850 Centennial Park Drive, Suite 301, Reston, VA 20191-1542, www.bia.org.

Ceramic Tile Institute Tile Manual (1991), Specification for Veneer Tile Installed on Exteriors in Portland Cement Mortar, Los Angeles: Ceramic Tile Institute.

Chrysler, John; Marble and Stone Slab Veneer, 3rd Edition. Masonry Institute of America. Torrance, CA, 2007. www.masonryinstitute.org.

Chrysler, John; Baltimore, Craig V.; Escobar, Thomas; Masonry Design Manual, Fourth Edition. Masonry Institute of America and International Code Council. Torrance, CA, 2007. www.masonryinstitute.org.

Curtain Wall Design and Construction (CDC), Dallas.

Design Guide for Anchored Brick Veneer Over Steel Studs, Western States Clay Products Association, www.wscpa.us.

International Conference of Building Officials (1994), Material, Testing and Installation Standards, 1994, Whittier: ICBO

National Concrete Masonry Association, NCMA, 13750 Sunrise Valley Drive, Herndon, VA 20171-4662, www.ncma.org.

NCMA TEK NOTE 3-6B, Concrete Masonry Veneers, www.ncma.org.

Van Houten, J. J. and Amrhein, J. E. (1977), Masonry Walls for Noise Control in Residential Construction, Los Angeles: Masonry Industry Advancement Committee.

INDEX

A

Acceptance----88
Adhered Veneer----29
  Adhered Veneer Installation----32
  Honeycomb Stone----31
  Terra Cotta----31
  Thin Brick----30
Adhered Veneer Installation Using Latex Portland Cement----69
Admixtures----14
Advantages of Thin Brick----30
Anchored Veneer----34
  Stone Veneer----37
Anchors----25
Applying Latex-Portland Cement Mortar----69
ASTM Stone Specifications----8

B

Brick Veneer Units----65
Brick Specifications for Latex-Portland Cement Mortar----70
Block Veneer, Concrete Units----43
Building Stone Surface Finish----9

C

Cements----13
  Masonry Cements----13
  Mortar Cements----13
  Portland Cement----13
Ceramic Veneer Units----91
Characteristics----11
Classification of Fabricated Stone for Soundness--10
Classification of Stone----8
  Geological Classification----8
  Texture of Quarried Stone----9
Clay Masonry Units----3
  Grade of Brick----4
  Visual Inspection----5
Cleaning----68, 76, 95
Cleaning and Protection----88
Codes and Standards----79
Color----15
Concrete Masonry Units----6
Connectors----24
  Anchors----25
  Wall Ties----24
Common Portland Cement-Lime Mortar Mixes----13
Compression Strength----6
Curing----95

D

Definitions----28, 70
Delivery, Storage and Handling----83
Disadvantages of Thin Brick----30

E

Expansion Joints----26
Expansion Joint Details----59

F

Fabrication----85

Finish----89
Finishes for Stone----87
Flashing----20
Floor Connection Details----54

G

General----3, 27, 45
Geological Classification----8
Glossary of Terms----97
Grade of Brick----4
Grade Recommendations for Face Exposures----4
Grout----74
Grouting of Brick Pavers----69
Guide Specification for Adhered Thin Clay Brick or Stone Slab Masonry Veneer----63
Guide Specification for Adhered Veneer Using Latex Portland Cement Mortar----69
Guide Specification for Anchored Brick Veneer----72
Guide Specification for Ceramic Veneer----89
Guide Specification for Exterior Stone Cladding----77
Guide Specification for Furnishing and Installing Ceramic Veneer----91
Guide Specification for Veneer Tile Installed on Exteriors in Portland Cement Mortar----94

H

Height Limit for Foundation----34
History of Masonry Veneer----1
Honeycomb Stone----31
Hydrated Lime----14

I

Inspection----89, 93
Inspection/Preparation----87
Installation----28, 67, 87, 92
Installation of Interior and Exterior Brick Paving----69
Introduction----1, 70

M

Masonry Cements----13
Masonry Units----3
  Clay Masonry Units----3
    Grade of Brick----4
    Visual Inspection----5
  Concrete Masonry Units----6
    Compression Strength----6
  Stone----7
    ASTM Stone Specifications----8
    Classification of Stone----8
      Classification of Fabricated Stone For Soundness----10
      Geological Classification----8
      Texture of Quarried Stone----9
  Terra Cotta----11
    Characteristics----11
Materials----3, 84, 94
Minimum Thickness of Weather Coverings----27
Mortar----12, 66, 74
  Mortar Materials----13
    Admixtures----14
    Cements----13
      Masonry Cements----13
      Mortar Cements----13
      Portland Cement----13
    Color----15
    Hydrated Lime----14
    Mortar Sand----14
    Water----14
  Pre–Blended Mortar----15
  Proportion Specification----13
Mortar and Grout----91
Mortar Cements----13
Mortar Proportions for the Unit Masonry----12
Mortar Sand----14

O

Other Materials----75

P

Package Labeling----71
Performance Criteria----80
Permissible Variations in Dimensions----5
Physical Properties----89
Physical Requirements----90
Plans and Specifications----78
Pointing----67
Portland Cement----13
Pre–Blended Mortar----15
Product Delivery, Storage and Handling----65, 74
Properties of Glazed and Vitrified Ceramic Finishes----89
Proportion Specification----13

Q

Qualifications----77
Quality Assurance----63, 73

R

References-----117
Repair of Stone-----88
Roof/Parapet Details-----45

S

Sand for Masonry Mortar-----14
Sampling and Testing and Procedures-----70
Schedule-----68
Sealants-----88
Sealing-----68, 76
Setting Adhered Veneer-----69
Setting Brick Paving-----69
Setting Drawings-----91
Shelf Angles-----19
Shelf Angle/Flashing Details-----48
Sill and Jamb Details-----51
Size and Shape-----89
Slab-Type Veneer-----41
Stone-----7
  ASTM Stone Specifications-----8
  Classification of Stone-----8
Stone Slab Units-----65
Stone Veneer-----37, 39
Strength and Absorption Requirements-----7
Strength of Adhered Veneer-----70
Structural Support Backup Materials-----17
  Steel Stud Backup-----18
  Structural Concrete Backup-----17
  Structural Masonry Backup-----17
  Wood Stud Backup-----18
Submittals-----79
Surface Preparation-----66
System Configuration-----61
System Detail Requirements-----27

T

Testing-----81
Terra Cotta-----11, 31, 42
  Characteristics-----11
Texture of Quarried Stone-----9
Thin Brick-----30
Tolerances-----86
Types of Mortar Joints-----15
Typical Details-----45

U

Unit Design-----89

V

Veneer Requirements-----27
  Definitions-----28
  Installation-----28
Veneer Units-----74
Visual Inspection-----5

W

Wall Base Details-----55
Wall Ties-----24
Warranty-----83
Water-----14
Weep Holes-----22
West Coast Veneer Stone-----10
Work Specified in Other Sections-----94
Workmanship-----75, 95
Wood Stud Backup-----18

Essential Codes and References From ICC

A

B

C

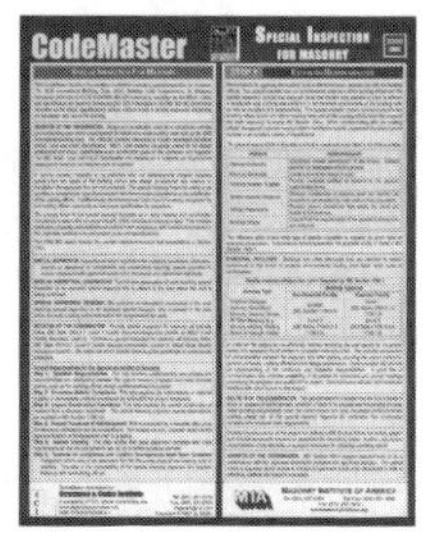

D

A: 2006 INTERNATIONAL BUILDING CODE®

This comprehensive code features time-tested safety concepts, structural, and fire and life safety provisions covering means of egress, interior finish requirements, comprehensive roof provisions, seismic engineering provisions, innovative construction technology, occupancy classifications, and the latest industry standards in material design. It is founded on broad-based principles that make possible the use of new materials and new building designs. The scope of this code covers all buildings except one- and two-family dwellings of three stories or fewer. (680 pages)

SOFT COVER #3000S06 | LIST $105 | **ICC MEMBER $31.50**

LOOSE LEAF #3000L06 | LIST $118 | **ICC MEMBER $88.50**

B: 2006 IBC® HANDBOOK: FIRE- AND LIFE-SAFETY PROVISIONS

Now published in full color, this updated bestseller offers expanded, completely detailed, and in-depth discussion on the provisions in Chapters 1–15 and 24–34 of the 2006 IBC®. The book is authored by ICC code expert Doug Thornburg, AIA, and contains more than 400 drawings and figures to clarify the application and intent of the code provisions. It also includes a CD-ROM containing the Handbook as well as ICC/ANSI A117.1-2003. An essential reference for every building official, fire marshal, architect and engineer. This exclusive publication comes in an 8½- by 11-inch, hard-cover format. (582 pages)

#4000S06 | LIST $93 | **ICC MEMBER $74**

C: 2006 INTERNATIONAL EXISTING BUILDING CODE®: CODE & COMMENTARY

This valuable tool contains the complete text of the IEBC® with tables and figures, corresponding commentary after each section, and suggestions for effective application. (482 pages)

#3560S06 | LIST $70 | **ICC MEMBER $56**

D: CODEMASTER: SPECIAL INSPECTION FOR MASONRY (2006 IBC®)

A four-page guide that provides inspectors with a step-by-step process to ensure masonry meets the requirements of the 2006 IBC®. The guide identifies six steps involved in masonry inspections and provides photos of materials, construction drawings and reference charts. Separate sections address materials compliance, construction conditions, reinforcement placement and grout inspection, quality assurance and correcting deficiencies.

#9350S06 | LIST $9 | **ICC MEMBER $7.50**

ORDER NOW! 1-800-786-4452 | www.iccsafe.org

8-61804-09